GLACIOTECTONIC LANDFORMS AND STRUCTURES

GLACIOLOGY AND QUATERNARY GEOLOGY

Series Editor:

C. R. BENTLEY

*University of Wisconsin-Madison,
Department of Geology and Geophysics,
Madison, Wisconsin, U.S.A.*

Glaciotectonic Landforms and Structures

by

JAMES S. ABER
Emporia State University, Kansas, U.S.A.

DAVID G. CROOT
Plymouth Polytechnic, Devon, U.K.

MARK M. FENTON
Alberta Research Council, Alberta, Canada

Kluwer Academic Publishers
Dordrecht / Boston / London

Library of Congress Cataloging-in-Publication Data

Aber, James S.
　Glaciotectonic landforms and structures / by James S. Aber, David
G. Croot, and Mark M. Fenton.
　　p.　cm. -- (Glaciology and quaternary geology)
　Bibliography: p.
　Includes index.
　ISBN 0-7923-0100-5
　1. Glacial landforms.　I. Croot, David G.　II. Fenton, Mark M.
III. Title.　IV. Series.
GB581.A24　1989
551.3'15--dc19   88-34047

ISBN 0-7923-0100-5

Published by Kluwer Academic Publishers,
P.O. Box 17, 3300 AA Dordrecht, The Netherlands

Kluwer Academic Publishers incorporates the publishing programmes
of D. Reidel, Martinus Nijhoff, Dr W. Junk and MTP Press.

Sold and distributed in the U.S.A. and Canada
by Kluwer Academic Publishers,
101 Philip Drive, Norwell, MA 02061, U.S.A.

In all other countries, sold and distributed
by Kluwer Academic Publishers Group,
P.O. Box 322, 3300 AH Dordrecht, The Netherlands

printed on acid free paper

All rights reserved
© 1989 by Kluwer Academic Publishers
No part of the material protected by this copyright notice may be reproduced or utilized in
any form or by any means, electronic or mechanical, including photocopying, recording, or
by any information storage and retrieval system, without written permission from the
copyright owners.

Printed in the Netherlands.

To our parents and families

TABLE OF CONTENTS

Acknowledgments ix

Chapter 1. Nature of glaciotectonism
  Historical development 1
  Definition of glaciotectonism 6
  Glaciotectonic structures 8
  Glaciotectonic landforms 9

Chapter 2. Hill-hole pair
  Introduction 13
  Wolf Lake, Alberta, Canada 16
  Herschel Island, Yukon, Canada 21
  Eyjabakkajökull, Iceland 24

Chapter 3. Large composite-ridges
  Introduction 29
  Møns Klint, Denmark 34
  Dirt Hills and Cactus Hills, Saskatchewan, Canada 37
  Prophets Mountains, North Dakota, United States 42

Chapter 4. Small composite-ridges
  Introduction 47
  Brandon Hills, Manitoba, Canada 49
  Utrecht Ridge, the Netherlands 54
  Vatnajökull, Iceland 59
  Spitsbergen, Svalbard, Norway 65

Chapter 5. Cupola-hills
  Introduction 71
  Ristinge Klint, Langeland, Denmark 73
  Gay Head, Martha's Vineyard, Massachusetts, United States 77
  Hvideklint, Møn, Denmark 83

Chapter 6. Megablocks and rafts

Introduction	91
Esterhazy, Saskatchewan, Canada	94
Southern Alberta, Canada	95
Kvarnby, Skåne, Sweden	98

Chapter 7. Diapirs, intrusions and wedges

Introduction	103
Kansas Drift, Kansas, United States	106
Herdla Moraines, Norway	110
Systofte, Falster, Denmark	116

Chapter 8. Applied glaciotectonics

Introduction	119
Highwall failure, Highvale Coal Mine, Alberta, Canada	121
Highway construction, Maymount, Saskatchewan, Canada	127
Diatomite quarries, Fur, Denmark	129

Chapter 9. Distribution of glaciotectonic phenomena

Continent-scale distribution	135
Regional distribution	142
Model for lobate pattern of glaciotectonism	148
Kineto-stratigraphy	149

Chapter 10. Dynamism of glaciotectonic deformation

Fundamental cause of glaciotectonism	155
Initiation of thrust faulting	159
Continuation of thrust faulting	162
Scale models of glaciotectonism	164

Chapter 11. Glaciotectonic analogs

Introduction	169
Mississippi Delta mudlumps, Mississippi, United States	169
Thin-skinned thrusting	173
Convergent plate boundary	179

Bibliography	183
Index	195

ACKNOWLEDGMENTS

Writing this book would hardly have been possible without the cooperation and encouragement of many people. We wish to particularly thank Professors Asger Berthelsen and Aleksis Dreimanis, who were instrumental in focusing scientific inquiry into glacial and glaciotectonic processes, deposits, landforms, structures and stratigraphy, and who were responsible for bringing the co-authors together.

We wish to additionally thank our many colleagues, who through discussions, encouragement, and sharing of information have directly or indirectly contributed to this book. These individuals include: Inge Aarseth, Lena Adrielsson, Harold V. Andersen, Lawrence D. Andriashek, John P. Bluemle, Earl A. Christiansen, Lee Clayton, Donald R. Coates, Louis F. Dellwig, Wakefield Dort, Jr., Lynda A. Dredge, Mike Garton, Karl Gripp, Sylvi Haldorsen, Michael Houmark-Nielsen, Rod A. Klassen, Rudy W. Klassen, Erik Lagerlund, Jan Lundqvist, Jan Mangerud, Rod A. McGinn, Steven R. Moran, Ernest H. Muller, Robert N. Oldale, John Pawlowicz, Stig A. Shack Pedersen, Kai Strand Petersen, Bertil Ringberg, Hanna Ruszczyńska-Szenajch, Steen Sjørring, Archie MacS. Stalker, James T. Teller, Sigurdur Thorarinsson, Dick F.M. van der Wateren, Donald Watkins, and M. William ter Wee.

During different stages of field work and preparation of the book manuscript the authors have received financial support from the following organizations: Alberta Research Council, Chadron State College, Commission for Educational Exchange between Denmark and the United States, Emporia State University, Icelandic National Research Council, Natural Research Council of United Kingdom, Natural Science Research Council of Sweden, Norwegian Marshall Fund, Plymouth Polytechnic, TransAlta Utilities Corporation, University of Bergen, University of Copenhagen, and University of Regina. We also wish to thank J. Pawlowicz, who prepared some of the figures.

CHAPTER 1

NATURE OF GLACIOTECTONISM

Historical Development

The glacial theory was born in the 1830s amid the splendor of the Swiss Alps. Jean de Charpentier is properly credited for initially developing the glacial theory, and Louis Agassiz popularized the new concept of former extension of glaciers and ice sheets (Teller 1983). Charpentier based his theory on distribution of three features: large erratic boulders, moraines, and abrasion marks on boulders and bedrock. In combination, these features could only be explained as the results of former glaciation. Glacial theory has from the beginning, thus, rested on two groups of geological field evidence: (1) features formed by glacial erosion and (2) features created by glacial deposition.

The possibility that glaciers could *deform* shallow crustal rocks and sediments was not recognized until a few decades later. Charles Lyell (1863) was apparently the first geologist to discuss the origin of contorted glacial strata at Norfolk, England, in the Italian Alps, and elsewhere. He suggested three possible mechanisms for deformation: (1) pushing by stranding icebergs, (2) melting of buried ice masses, and most importantly (3) pushing before advancing glacier ice. Lyell's *The Antiquity of Man* (fig. 1-1) was widely read, and his comments on glacially deformed sediments must have influenced many geologists.

During the following two decades, ice-pushed structures were recognized in several now classic locations: Skåne, southernmost Sweden (Torell 1872, 1873; Erdmann 1873); Møns Klint, Denmark (fig. 1-2) and Rügen, German Democratic Republic (Johnstrup 1874); and southern New England islands in the United States (Merrill 1886a). The obvious structural disturbances at these places had previously been ascribed to a variety of causes: landslides, volcanism/intrusion, mountain building, *etc.*

Recognition of ice-shoved features from interior continental locations did not come about until early in this century. Various glaciotectonic structures were described near Minneapolis, Minnesota (Sardeson 1905, 1906) and in Poland (Lewiński and Różycki 1929). Hopkins' (1923) analysis of large ice-shoved hills in eastern Alberta is perhaps one of the best early studies. He emphasized the similarity in structural style of these hills compared to the foothills of the Canadian Rocky Mountains.

The first geologist to really specialize in the study of glacially deformed structures was George Slater, an Englishman, who chose the subject of ice-push deformation for his doctoral dissertation at the University of London. He studied ice-shoved hills in England (Slater 1927a, 1927b), Denmark (Slater 1927c, 1927d),

> THE GEOLOGICAL EVIDENCES
>
> OF
>
> # THE ANTIQUITY OF MAN
>
> WITH REMARKS ON THEORIES OF
>
> THE ORIGIN OF SPECIES BY VARIATION
>
> By SIR CHARLES LYELL, F.R.S.
>
> AUTHOR OF 'PRINCIPLES OF GEOLOGY,' 'ELEMENTS OF GEOLOGY,' ETC. ETC.
>
> *THIRD EDITION, REVISED*
>
> ILLUSTRATED BY WOODCUTS
>
> LONDON
> JOHN MURRAY, ALBEMARLE STREET
> 1863

Fig. 1-1. Cover page of Lyell's *The Antiquity of Man* (1863), which contains the first general discussion of contorted drift.

Canada (Slater 1927e), the United States (Slater 1929), and the Isle of Man (Slater 1931). He was the first to use the term *glacial tectonics* (Slater 1926), which is now generally shortened to *glaciotectonics* (American) or *glacitectonics* (British).

Unfortunately, the Danish geologist Axel Jessen, who had once helped Slater, later accused Slater in no uncertain terms of plagiarizing and misrepresenting his (Jessen's) work at Lønstrup Klint, Denmark (Jessen 1931). In spite of Slater's prolific publications, glaciotectonism was still considered an unusual or peculiar manifestation of the glacial theory. Small contortions in glacial strata were commonly recognized, but many geologists continued to deny that glaciation could create large deformations involving much bedrock.

Fig. 1–2. Earliest known illustration of Møns Klint, from Pontoppidan's *Danske Atlas* (1764). Chalk mass in center, Sommerspiret (B), stands in a vertical position with the pinnacle 102 m above sea level.

The growth of glaciotectonic research in Europe during the middle portion of this century was checkered by distinct national differences. The study of glaciotectonic phenomena emerged briefly as a specific field of research in Denmark during the 1930s under the leadership of Helge Gry (Univ. Copenhagen) and Jessen (Geological Survey Denmark).

Gry's (1940) analysis of ice-shoved hills in the Limfjord region of northwestern Denmark demonstrated the full potential of combining structural geology with glacial stratigraphy and geomorphology. Gry was strongly influenced by observations of neoglacial push-moraines on Spitsbergen made by the German geologist Gripp (1929). However, when Gry moved to a position at the Geological Survey, teaching of glaciotectonics ceased at the University, and the study of Danish glaciotectonics languished for many years thereafter.

Glaciotectonics next emerged as an important subject of research in the Netherlands during the 1950s (de Jong 1952; Maarleveld 1953). The reason for Dutch interest in glaciotectonics is obvious: ice-shoved ridges are the most conspicuous topographic features in a country otherwise noted for its flatness. The Netherlands continues to be a center for glaciotectonic research, as shown by numerous detailed investigations (Ruegg and Zandstra 1981; van der Meer 1987).

Glaciotectonics also became a significant field of research in Poland (Jahn 1956; Dylik 1961; Galon 1961) during this period. Glaciotectonics is perhaps the most active research field in glacial geology today in Poland (Ruszczyńska-Szenajch 1985), where ice-pushed structures are increasingly important for overall Quaternary geology (Brodzikowski and van Loon 1985). Meanwhile, geological consideration of German glaciotectonic features was also developing. Recent research in the northern German Federal Republic reflects a renewed interest in ice-push deformation there (Stephan 1985; van der Meer 1987).

Glaciotectonic research in Denmark was revived during the 1970s by Asger Berthelsen (Univ. Copenhagen), who applied his experience with hard-rock structural geology to unraveling glaciotectonic phenomena. He developed the method of *kineto-stratigraphy* (Berthelsen 1973, 1978:25), in which the 'main emphasis is placed on the study of the directional elements that reflect the movement patterns (kinetics) of former ice sheets.' Berthelsen motivated students, and his kineto-stratigraphic method was highly successful for working out Weichselian glacial stratigraphy in Denmark. In fact, glaciotectonic analysis is now an integral part of geological mapping in Denmark (Petersen 1978).

Following Slater's controversial career with glaciotectonics, British interest in the subject was minimal. The classic chalk rafts and contorted drift along the Norfolk coast received some attention in the 1960s (Peake and Hancock 1961; Harland et al. 1966), and a revival in glaciotectonic studies there was led by Peter H. Banham (Univ. London). Banham (1975, 1977) also applied hard-rock geologic principles to interpretation of glaciotectonic structures. Banham's work complemented that of Berthelsen along with the Soviet geologist Lavrushin (1971). Together, they built the methodologic foundation for modern glaciotectonic research.

Aside from several early studies on southern New England islands (Hollick 1894; Woodworth 1897; Upham 1899) and a few other isolated investigations, little glaciotectonic research was carried out in North America until the late 1950s. Part of the explanation may be the fact that fewer glacial scientists were faced with a much larger and less developed geographic area to investigate.

The glaciated region covered by the Laurentide Ice Sheet in North America is approximately twice that of combined Fennoscandian/British Ice Sheet coverage (Flint 1971). Beginning in the 1950s, large-scale topographic maps and aerial photographs became widely available, which greatly facilitated geologic reconnaissance. At the same time, many states/provinces commenced systematic geologic mapping programs.

The North American glaciotectonic renaissance began in western Canada on two fronts. Working in southern Saskatchewan, A.R. Byers (Univ. Saskatchewan) demonstrated the ice-thrust genesis of the Dirt Hills and Cactus Hills. These are among the largest and best-developed glaciotectonic hills in the world (Byers 1959). Walter Kupsch (Univ. Saskatchewan) expanded on Byers' work to include similar large ice-shoved hills across southern Saskatchewan and eastern Alberta

(Kupsch 1962). At the Saskatchewan Research Council, reconnaissance mapping, done mainly by Earl Christiansen, resulted in much additional knowledge of glaciotectonic features (Christiansen 1961, 1971a, 1971b; Parizek 1964; Christiansen and Whitaker 1976).

At the same time, equally impressive ice-shoved ridges on the Yukon Coastal Plain were described by J.R. Mackay (1959). These features, developed wholly within permafrost, were the basis for Mackay and colleague W.H. Mathews (both Univ. British Columbia) to develop a general theory for glaciotectonic thrusting (Mathews and Mackay 1960; Mackay and Mathews 1964). They were strongly influenced by Hubbert and Rubey's (1959) analysis of overthrust faulting in mountains.

The Saskatchewan discoveries soon spilled over into North Dakota, where a county mapping program was underway. Led by Lee Clayton (Univ. North Dakota), Steven R. Moran and John L. Bluemle (both North Dakota Geological Survey), the state became a productive center for glaciotectonic research during the 1970s. Many local studies were conducted, culminating in a new *Geologic Map of North Dakota* (Clayton et al. 1980), on which a variety of glaciotectonic landforms are shown.

Theoretical analysis accompanied the field mapping (Moran 1971; Clayton and Moran 1974; Bluemle and Clayton 1984). When Moran moved to the Alberta Research Council, the geographic area was expanded to include the entire glaciated Great Plains region of the United States and Canada. The result was the first attempt at continent-scale synthesis concerning ice-sheet dynamics, distribution of glacial landforms, and genesis of glaciotectonic phenomena (Moran et al. 1980). Clayton later moved to the Wisconsin Geological Survey, with the unsurprising result that glaciotectonic features are now recognized in that state.

With increasing surface and subsurface information, it now appears that glaciotectonic phenomena are omnipresent in glaciated regions underlain by sedimentary bedrock or thick drift (Moran et al. 1980) and are even common in thin drift resting on crystalline bedrock. A variety of distinctive landforms is now attributed either wholly or partly to glaciotectonic genesis. Hence, glaciotectonic features must be included with depositional and erosional features as primary field

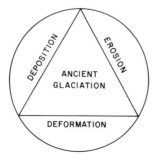

Fig. 1–3. Triad of effects created by glaciation, on which modern glacial theory is based.

evidence for glaciation. The modern glacial theory is, therefore, supported by a triad of field evidence, including erosional, deformational, and depositional features (fig. 1-3).

International recognition of the significance of glaciotectonic phenomena came in 1982, when a Work Group on Glacial Tectonics (WGGT) was established within the International Union for Quaternary Research (INQUA). A. Berthelsen organized WGGT and served as its first President. The overall goals of WGGT are: to initiate and stimulate studies of glaciotectonic phenomena in both recent and ancient glacial environments, to promote interdisciplinary collaboration between scientists working in different parts of the field, and to increase glaciotectonic curriculum in academic teaching and professional training (WGGT Newsletter, 3/1987).

The importance of glaciotectonic structures and landforms is now well accepted among many researchers actively engaged with studies of glacial geology and geomorphology. Special symposia and field conferences have been conducted (Sjørring 1985; van der Meer 1987; Croot 1988a), and glaciotectonic examples have been used for laboratory exercises in structural geology (Aber 1988a).

Little of this new emphasis has spread into general geology, however, or to the greater sphere of natural science. One problem is the highly fragmented publication of scientific reports dealing with the subject (Aber 1988e): diverse journals and government documents, various languages and countries, and different geologic, geographic, glaciologic, and engineering specialties.

The teaching of glaciotectonics has not kept up with research developments. As an example, Flint's (1971) comprehensive text book devotes five chapters to glacial erosion, deposition and landforms, of which only four pages are given to ice-thrust features. No criticism of Flint's book is implied, for its coverage of glaciotectonic phenomena is actually better than several popular text books published since.

In one recent text book on glacial processes by Drewry (1986), 'glaci-tectonic' structures are briefly discussed as deformations within glacier ice. However, deformation of substratum material and creation of conspicuous ice-shoved landforms are not even mentioned. A void clearly exists at present between research and teaching of glacial theory, a void which this book is intended to fill.

Definition of Glaciotectonism

Considerable uncertainty surrounds the exact meaning and use of the term glaciotectonic, because various deformed structures are common both in glacier ice and in glacial deposits. Although Slater (1926) did not define the phrase glacial tectonics, he used it in reference to structural disturbances in both drift and bedrock as well as deformations in glacier ice. Similar, if somewhat vague, references have been given by other geologists in the years since.

An important review by Occhietti (1973) includes five categories of glacially related deformation:

(1) *glaciotectonic* – deformation of pre-existing substratum (drift and bedrock) by active ice movement.
(2) *glaciodynamic* – primary structures (such as till fabric) produced within ground moraine by active ice.
(3) *glaciostatic* – deformation of ground moraine and substratum by static ice loading.
(4) *glaciokarstic* – deformation accompanying freezing/thawing of buried dead-ice masses.
(5) *iceberg drifting* – deformation of sea or lake sediments by grounded icebergs.

The latter two categories may be eliminated from further consideration, leaving three: glaciotectonic, glaciodynamic, and glaciostatic. Glaciodynamic structures are primary features produced in till during initial deposition of the sediment. Such structures may be of great interest for interpreting the direction of ice movement or the physical conditions at the base of the glacier. However, as primary features, they cannot be considered as disturbances or deformations in the normal sense of structural geology (Billings 1972). Such glaciodynamic features, labeled *endiamict* structures by Banham (1975), are adequately dealt with elsewhere and will not be discussed in detail here.

A glacier or ice-sheet may induce deformations of pre-existing substratum material as a result of its forward (dynamic) movement or its vertical (static) loading. Both causes of deformation operate simultaneously, and the effects of each cannot usually be separated. Therefore, glaciotectonic and glaciostatic structures are considered as joint manifestations of secondary deformations produced during glaciation. The term *exodiamict* has been used for such secondary structures within the substratum (Banham 1975). Exodiamict structures and their landforms are the primary subject of this book.

Glaciotectonism may be defined as glacially induced structural deformation of bedrock and/or drift masses as a direct result of glacier-ice movement or loading (based on Moran 1971; Aber 1985a). In English, the word *glaciotectonic* is an adjective that should be used to modify a noun, for example glaciotectonic landform. In noun forms, *glaciotectonics* refers to features or results of glaciotectonic deformation, whereas *glaciotectonism* refers to the processes of glaciotectonic deformation.

Glaciotectonics excludes: deformed structures within glacier ice, structures falling in categories 2, 4 and 5 above, and other crustal structures not created by active glacier ice. Depression and rebound of the lithosphere as a consequence of glaciation and deglaciation is a form of glaciotectonism, long recognized and much studied (Andrews 1970; Flint 1971). However, the emphasis in this book is placed on shallow (< 200 m deep) structures and landforms. Lithospheric depression is considered here only insofar as it may bear on the genesis of shallow glaciotectonic features.

The identification of glaciotectonic structures and landforms is based on two fundamental criteria: (1) presence of recognizable masses of pre-existing bedrock and/or drift and (2) presence of glacially induced deformations within those masses. Bedrock is used here in a general sense, as any pre-Quaternary rock or sediment material against or over which the glacier moved. The original bedding, lithology, or other primary characteristics must be clearly visible in the material.

Deformation, in the form of folds, faults, breccia, slickensides, or other disturbances, may be produced essentially *in situ* (autochthonous) or during the transportation and deposition of a detached mass (allochthonous). Such deformations must be the result of glacially imposed stresses and are not merely pre-existing structures inherited from the parent material.

Glaciotectonism can be viewed in some cases as a kind of incomplete or partial glacial erosion of bedrock. Conversely, it can also be considered as a form of glacial deposition in other cases. However, neither approach is entirely satisfactory (Ruszczyńska-Szenajch 1988); glaciotectonic features represent those situations where glacier ice was capable of deforming pre-existing strata without completely removing or destroying the rock or sediment beyond recognition. Whether glaciotectonic features would develop was controlled to a large extent by the physical conditions of the substratum material. The same dynamic conditions that elsewhere caused conventional glacial erosion and deposition created glaciotectonic features in appropriate kinds of bedrock or drift.

Glaciotectonic Structures

Glaciotectonic structures range in scale from microscopic to continental. Deformed material includes mostly sedimentary strata varying from unconsolidated to poorly and moderately consolidated in character. Less commonly, well-consolidated sedimentary strata and even crystalline rocks have also suffered glaciotectonic deformation (Kupsch 1955; Babcock *et al.* 1978). It should be understood,

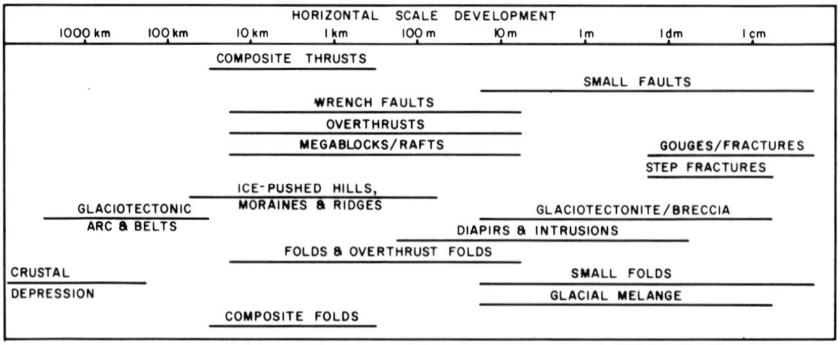

Fig. 1–4. Chart of common glaciotectonic structures arranged according to their typical horizontal scales (logarithmic). Ductile structures toward bottom; brittle structures toward top. Based on Occhietti (1973, Tab. II).

however, that the conditions existing at the instant of deformation may have been quite different from what we observe today. For example, many deformations took place under conditions of high confining pressure or while the sediment mass was permafrozen.

A complete listing of all glaciotectonic structures is probably not possible, simply because so much variety exists in type, style and size. Nonetheless, several common glaciotectonic structures can be recognized (fig. 1–4). In a general way, glaciotectonic structures can be divided in two broad categories – ductile and brittle – depending on the nature of deformation.

Ductile deformation takes place by internal creep or flow of material in a plastic or fluid manner. During ductile deformation, the rock mass has essentially no internal strength, so that very small pressure differences may result in substantial changes in the size or shape of the mass. Ductile structures are most typical of unconsolidated or fine-grained strata, such as clay, silt, shale or chalk, that were deformed under high confining pressures. Various folds, intrusions, diapirs, and contortions represent ductile deformation.

Brittle deformation results when rock masses fail by fracturing along discrete planes. Deformation is accomplished by movement or adjustment along fracture planes, whereas the internal fabric of the rock between fractures shows no permanent deformation. Brittle structures are most characteristic of consolidated or coarse-grained strata, such as sand, gravel, sandstone or limestone, that were deformed under low confining pressure. Joints, faults, breccia, fissures and other fractured structures exhibit brittle deformation.

Ductile and brittle structures are often intimately associated within the same sequence of deformed strata. This is because strength varies for each rock layer depending mainly on lithology and thickness of the layer. Within a stratified sequence, layers of differing lithology and thickness each respond to deforming pressure differently. As a result, some layers develop brittle structures, and others show ductile deformation (Plate I). Also, a particular rock that displays brittle deformation in some places may show ductile structures elsewhere due to variations in pressure, temperature or fluid content during deformation.

Many scientists have pointed out the striking similarity in deformational style between ice-shoved hills and true mountains (Hopkins 1923; Berthelsen 1979; van der Wateren 1985; Banham 1988a). All manner of hard-rock structures described from mountains and shields have been recognized in glaciotectonic settings (Banham 1977). Glaciotectonic structures often mimic those seen in igneous and metamorphic rocks. In fact, ice-shoved hills may be regarded as natural scale-models of mountains; the only significant difference is size (Croot 1987).

Glaciotectonic Landforms

Glaciotectonic landforms are the surface or morphologic expressions of subsurface structures resulting from glacial deformation of soft bedrock and/or drift. The landforms may display their original or primary morphology as initially created during glaciation. This is particularly true of young (late Wisconsin/Weichselian) features that were little modified by subsequent glaciation or by postglacial erosion and deposition. In such cases, the morphology may be a direct expression of subsurface structures.

In other situations, glaciotectonic landforms were significantly altered by later glacial or nonglacial events. The present landforms may be little more than erosional ruins of the original forms or may be largely hidden beneath younger deposits. In these areas, the morphology is only a subdued reflection of subsurface structures. In all cases, some knowledge of subsurface structure and stratigraphy is invaluable for properly interpreting the landforms.

Glaciotectonic landforms comprise a variety of types, including hills, ridges, buttes and plains, all of which are constructed wholly or partly of pre-existing bedrock and/or drift masses. Many kinds of glaciotectonic landforms have been described during the past century. A great many terms have been used for ice-shoved hills, including (Kupsch 1962): *Stauchrücken, Stauchmoränen* and *Stauchendmoränen* (German); *stuuwmorenen* and *stuuwwallen* (Dutch); pseudo-moraine, push moraine, thrust moraine, ice-pushed ridges, *etc.* (English).

The general term *ice-shoved hill* is used here to refer to any glaciotectonic landform of constructional nature. Depressions or basins formed by glacial erosion will be considered insofar as they may relate to ice-shoved hills. The term *push-moraine* is restricted here to those ice-shoved hills that are composed largely or wholly of deformed glaciogenic strata. The terms floe, raft, scale, and megablock have all been used for the individual dislocated masses of bedrock and drift that make up ice-shoved hills. *Floe* is a general term for any kind of dislocated and deformed mass, and *scales* are blocks thrust into an imbricated or overlapping position. The terms *raft* and *megablock* both refer to large, comparatively thin masses lying in more-or-less horizontal positions.

Classification and regional mapping of these landforms has only been undertaken recently (Moran *et al.* 1980). Clayton *et al.* (1980) classified ice-shoved hills of North Dakota into three types: hill-depression forms, transverse-ridge forms, and irregular forms. This classification is significant, because it was the first attempt to classify glaciotectonic landforms for a large region, and because the hill-depression form was recognized as the fundamental type of glaciotectonic landform. A more elaborate classification is possible based on subdivision of these three types, inclusion of other landforms, and consideration of building materials.

An expanded classification for constructional glaciotectonic landforms includes five types (Aber 1988b): (1) *hill-hole pair*, (2) *large composite-ridges*, (3) *small composite-ridges*, (4) *cupola-hill*, and (5) flat-lying *megablock*. Each class represents an ideal genetic type within a continuous spectrum of glaciotectonic landforms (Table 1–1).

Table 1-1. Basic characteristics of constructional glaciotectonic landforms arranged in order of decreasing topographic prominence from the top down. See chapters 2-6 for further discussion.

Landform	Height (m)	Area (km^2)	Primary Material	Primary Morphology
Large composite-ridges	100 - 200	20 - >100	bedrock	subparallel ridge and valley system arcuate in plan
Hill-hole pair	20 - 200	<1 - >100	variable	ridged hill associated with source depression
Small composite-ridges	20 - <100	1 - >100	Quaternary strata/drift	subparallel ridge and valley system arcuate in plan
Cupola-hill	20 - >100	1 - 100	variable	smoothed dome to elongated drumlin with till cover
Megablock	0 - <30	<1 - 1000	bedrock	often concealed, flat buttes or irregular hills

Intermediate, transitional, or mixed landforms exist between these ideal types. For example, Herschel Island (Chapter 2) is the hill portion of a large hill-hole pair. Part of the island displays large composite-ridges, whereas the rest of the island resembles a cupola-hill. Such combination landforms are relatively frequent. Nonetheless, the five ideal types are distinctive enough to justify special recognition of each as an end-member class.

The materials of which glaciotectonic landforms are built may be divided into three categories: (1) pre-Quaternary bedrock that is usually, but not always, consolidated to some extent; (2) pre-existing Quaternary strata, both glacial and nonglacial, that are usually unconsolidated; and (3) contemporaneous drift, deposited and subsequently deformed during the same glaciation, that is always unconsolidated. Here, as with the landform types, these categories represent end members with many landforms being composed of more than one type of material. Regardless of their present state of consolidation, any of these materials may have been permafrozen at the time of deformation.

The case-history method of teaching, which is well established in law, medicine, engineering, and other fields, is adopted for this book. In the following chapters, actual case examples are selected to illustrate the typical or salient characteristics of each landform or structure type. These examples are representative of glaciotectonic analysis and demonstrate both its possibilities and limitations.

Some of these examples are readily accessible, others are remotely located; some are thoroughly investigated, others have received only cursory study; a few are famous for their scenic beauty (see Maps 1 and 2 for case-example locations). In any event, they collectively display structures and morphologic features that represent the spectrum of glaciotectonic phenomena.

CHAPTER 2

HILL-HOLE PAIR

Introduction

The hill-hole pair is perhaps the simplest and most instructive type of glaciotectonic landform. It represents a basic combination of ice-scooped basin and ice-shoved hill. Other types of glaciotectonic landforms are variations of the hill-hole theme. The association of individual ice-shoved hills with discrete source depressions was first described by Jessen (1931) from northern Jylland, Denmark.

The pairing of anomalous hills and upglacier depressions in central North Dakota was noted by Bluemle (1970) and Clayton and Moran (1974), who correctly recognized the glaciotectonic origin of the hill-hole pairs. In the past, these hills were often misidentified as kames or in-place outliers of bedrock, depending on their internal composition. Hill-hole pairs are now widely recognized.

Bluemle and Clayton (1984:284) described the hill-hole pair as, 'a discrete hill of ice-thrust material, often slightly crumpled, situated a short distance downglacier from a depression of similar size and shape.' The hill and associated depression are usually next to each other, but may be separated in some instances by as much as 5 km. Both pre-existing drift or bedrock may be involved in the dislocated hills.

The depression represents the source of material now in the hill. Depressions today are often the sites of bogs, lakes, estuaries, bays, or simply low spots in the surrounding topography. The volume of the depression should ideally be equal to or slightly less than the volume of the hill. Where this can be shown, a direct genetic link is demonstrated between the hill and hole. However, as depressions are often partly filled with younger sediment, it may be difficult to determine a depression's original volume.

In some cases, the source depression for a hill cannot be identified. The depression may actually exist but is hidden by younger sediment cover or is under a large lake or sea. It can be demonstrated in a few situations that the dislocated mass was a pre-existing hill that was simply moved to a new location (Moran *et al.* 1980). On the other hand, anomalous depressions without associated hills are also known in several regions, for example in central Poland (Ruszczyńska-Szenajch 1976, 1978) and eastern Alberta (Andriashek and Fenton 1988).

The basic morphology of typical hill-hole pairs is shown clearly by Antelope Hills, central North Dakota (fig. 2–1) and at Hundborg, western Denmark (fig. 2–2). Antelope Hills is composed mainly of deformed, soft Cretaceous or Paleocene sedimentary bedrock upthrust some 90-100 m above the surrounding landscape (Carlson and Freers 1975; Clayton *et al.* 1980). A depression to the west, partly occupied by a small lake at 1525 feet elevation, is the source for Antelope

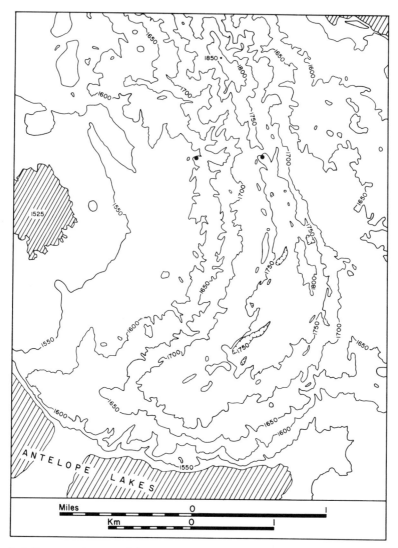

Fig. 2–1. Topographic map of Antelope Hills vicinity, central North Dakota. Elevations in feet; 50-foot contour interval (= app. 15 m). Perennial lakes shown by diagonal lining; locations of road cuts revealing deformed bedrock shown by solid dots.

Hills. The hill at Hundborg is situated along the Paleocene outcrop belt of northern Jylland, wherein spectacular ice-pushed structures were created in the soft clayey bedrock (Gry 1940). Sjørring Sø is the source depression for the hill at Hundborg. Characteristic morphologic features of simple hill-hole pairs include: 1. Arcuate or crescentic outline of hill; concave on the upglacier side, convex in the downglacier direction; 2. Multiple, subparallel, narrow ridges separated by equally narrow valleys following the overall arcuate trend of the hill; 3. Asymmetric cross profile of hill; higher with steeper slopes on convex (downglacier) side, lower with gentler

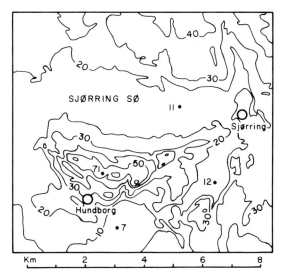

Fig. 2–2. Topographic map of Hundborg vicinity, western Denmark. Elevations in m; contour interval = 10 m.

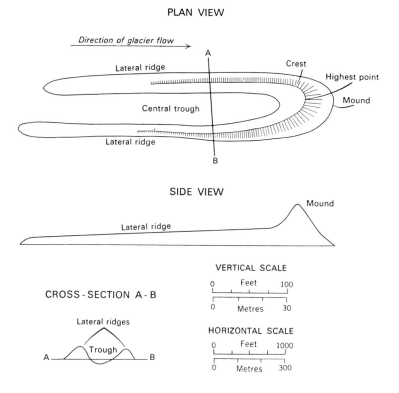

Fig. 2–3. Plan and side views and cross section of a typical murdlin from central Alberta. Taken from Stalker (1973a, fig. 6).

slopes on concave side; 4. Topographic depression on concave (upglacier) side of hill; area and shape of depression approximately equal to that of hill.

Where all these morphologic traits are present, a hill-hole pair may be identified with confidence, even without knowledge of subsurface conditions. However, original hill-hole morphology was commonly modified by later glacial and nonglacial processes, so the ideal set of characteristics is not always present. The sizes of hills and related depressions usually vary from 1 km^2 to 100 km^2, and often many hill-hole pairs of different sizes are found in close proximity to each other. Topographic and structural relief of hill-hole pairs generally ranges from 30 m to 200 m, although exceptions to these size and relief ranges are known.

A special variety of hill-hole pair consisting of an elongated loop with a central trough is found in central Alberta. Stalker (1973a) coined the term *murdlin* for this landform, because the hill resembles a drumlin when viewed from the side, but has its highest crest at the distal end and has a longitudinal depression (fig. 2–3). The lateral ridges and distal mound are formed of debris shoved from the trough. The origin of murdlins is obscure, but Stalker (1973a) believed they formed during the final stages of glaciation in places where a narrow tongue of active ice pushed through a marginal belt of dead ice.

The following case examples demonstrate hill-hole pairs in various settings. Wolf Lake, Alberta and Herschel Island, Yukon are both large Pleistocene hill-hole pairs. Herschel Island is constructed of permafrozen, preglacial Quaternary strata, whereas the internal nature of the hill at Wolf Lake is unknown. The smaller hill-hole pairs from Iceland were created in unfrozen sediments during Holocene ice advances and demonstrate the difficulty of determining hill-hole volume relationships.

Wolf Lake, Alberta, Canada

Wolf Lake is situated in east-central Alberta near the Saskatchewan border (fig. 2–4). A large hill of ice-shoved material, here called Wolf Hill, is located immediately south of the lake. This hill reaches a maximum elevation of 759 m (2476 feet), some 158 m above the normal level of Wolf Lake. To the west of Wolf Lake, smaller, streamlined, ice-shoved hills, drumlins, and flutes are well developed. The land surface is completely mantled by drift. Wolf Lake and Wolf Hill in combination, thus, represent a large hill-hole pair in a complex setting (Fenton and Andriashek 1983; Andriashek and Fenton 1988). The distinct morphologic expression alone makes this an outstanding example, even though almost nothing is known concerning its internal structures or materials.

Wolf Lake and Wolf Hill are both shaped approximately as similar, aligned parallelograms, 7.5 km long and 3 km wide. The alignment of the lake and hill is emphasized by a conspicuous, southwest-trending lineament formed by the straight

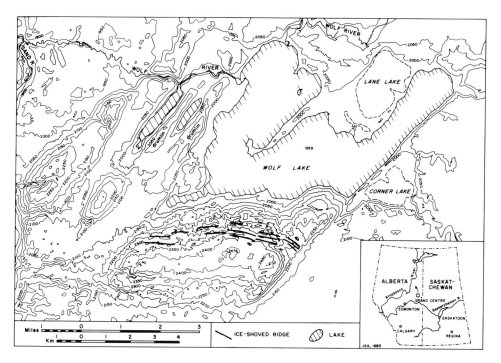

Fig. 2–4. Topographic map of Wolf Lake vicinity, Alberta. Position of ice-shoved ridges based on analysis of aerial photographs. Elevations in feet; contour interval = 50 feet (app. 15 m); perennial lakes shown by diagonal lining; ephemeral lakes shown by dashed outlines; F = fire watch tower.

eastern edge of Wolf Lake and the straight eastern flank of Wolf Hill (fig. 2–5). The Wolf Lake lineament is more than 11 km long and displays 135 m of total topographic relief.

Any lineament of similar prominence in a nonglacial setting would be recognized as a major vertical fracture, such as a strike-slip fault. This is the interpretation Andriashek and Fenton (1988) have given to the eastern as well as western boundaries of Wolf Lake (fig. 2–6). The boundaries of the ice-scooped depression are essentially tear faults along which material from Wolf Lake basin was shoved into Wolf Hill.

Wolf Hill consists of a single, steep-sided, east-west trending, asymmetric mound with the southern flank being steeper than the northern one and a rounded crest (fig. 2–6). A belt of narrow, parallel ridges occupies the hill's northern flank and aerial photographs show traces of these ridges extending down the eastern and western sides into the subsurface. Such ridges are the most typical morphologic trait of ice-shoved hills. The ridges indicate that at least the northern portion of the hill probably consists of imbricately stacked thrust blocks.

Wolf Lake partially fills the depression excavated during deformation. Scarps, about 10 m high, form the eastern, western and northern margins of the lake

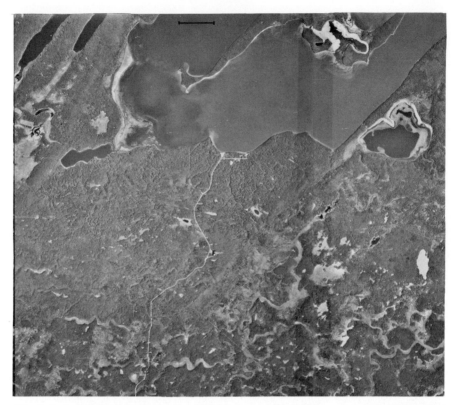

Fig. 2–5. Aerial photograph showing Wolf Lake and Wolf Hill, east-central Alberta. Note the lineament formed by the eastern edge of Wolf Lake and the eastern flank of Wolf Hill. Scale bar is 1 km long; north toward top; F = fire watch tower. Aerial photographs: A24452–15 and 16; copyright 1976. Her Majesty the Queen in Right of Canada, reproduced from the collection of the National Air Photo Library with permission of Energy, Mines and Resources Canada.

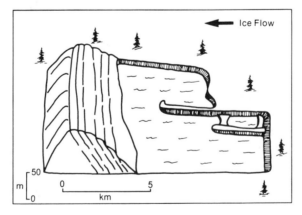

Fig. 2–6. Schematic diagram showing fault-bounded depression of Wolf Lake from which material was displaced into the ice-shoved hill. Ice flow from northeast toward southwest. Diagram from Andriashek and Fenton (1988).

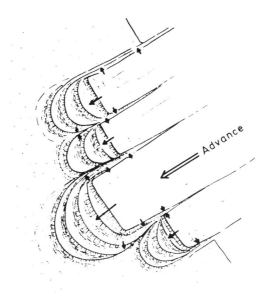

Fig. 2–7. Schematic map of block-movement model for glacier pushing. Long arrows show direction of hill-hole shoving; short arrows show direction of pushing for parallel ridges. Taken from Stephan (1985, fig. 3).

(Fenton and Andriashek 1983). Bathymetric maps show that the lake occupies three depressions corresponding to the eastern, central and western arms of the lake. Lane Lake is only a minor shallow marsh. The volume of the entire depression is only about 80% of the volume of Wolf Hill. This is believed to be the result of later infilling of the western and central arms of the lake. Corner Lake is a steep-sided depression lacking a hill on the downglacier side and is a 'hill-less hole'. The undisturbed terrain surrounding the hill-hole pair is essentially flat.

The model for glaciotectonic pushing by block movement of the glacier front, that was developed in northern Germany (Stephan 1985), may be applicable to Wolf Lake (fig. 2–7). Pushing of large masses occurred in front of individual ice blocks, whereas smaller ridges were pushed parallel to block side margins, which could also be the sites of strike-slip faulting. An *en echelon* pattern of hill-hole pairs connected at right angles by tear faults or elongated drumlins is the ideal result.

In the Wolf Lake vicinity, a series of ice blocks may have been responsible for scooping the following depressions: (1) Corner Lake, (2) eastern arm of Wolf Lake, (3) central Wolf Lake, (4) western arm of Wolf Lake, (5 and 6) smaller and larger finger lakes west of Wolf Lake. These depressions are bounded by tear faults, low ridges, or elongated drumlins. Wolf Hill was likely constructed by three of these ice blocks (2, 3 and 4). Aerial photographs of Wolf Hill do not, however, show that it is divided into three segments. Perhaps the contacts between segments are covered by sediment of englacial or superglacial origin. In any case, the morphologic expression of Wolf Lake and Wolf Hill demonstrates a clear genetic pairing of the two.

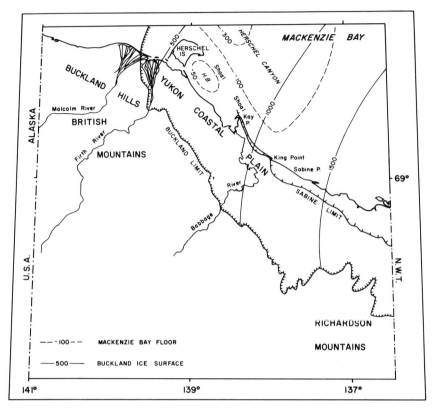

Fig. 2–8. Map of northern Yukon, Canada showing geographic and glacial features. Contours in feet (100 feet = app. 30 m); H.B. = Herschel Basin. Adapted from Mackay (1959, fig. 1) and Rampton (1982, figs. 1 and 18).

Morphologic and stratigraphic data have shown that all the features in the Wolf Lake area were formed by the southwest-flowing Primrose Lobe during the last, local phase of the Late Wisconsin Cold Lake glaciation. The small lakes and hills west of Wolf Lake show evidence of smoothing by glacial flow; however, Wolf Hill, the major feature, has a fresh and unmodified appearance. The nearest ice margin, for which there is evidence, lies about 4 km west and 10 km south of Wolf Lake (Andriashek and Fenton 1988).

Moran *et al.* (1980) concluded that Wolf Hill and other ice-shoved hills on the Great Plains of central North America were created in a narrow (2–3 km wide) frozen-bed zone at the margin of active ice. Meanwhile, streamlined morphology was molded under a thawed bed farther upglacier. In the Wolf Lake area, the thrusting may have taken place during a stillstand or minor readvance to the previously mentioned ice margin with the local molding of the thrust sediment taking place as the frozen-bed zone moved through the area. Clearly the thrusting is a late-phase phenomena, because Wolf Hill shows no evidence of prolonged sculpting by overriding ice.

Area	Herschel Island	Malcolm Lake to Babbage River	Erosion Surface	Kay Point and southeast	King Point	Sabine Point and northwest
General Stratigraphic Descriptions	Marine clays; possibly contain sequence of freshwater sediments. Vegetation--- shrub tundra. --- 'Mixed' sediments from shallow marine and brackish environments. Vegetation--- boreal forest. --- Marine clays	Marine clays --- Deformed sands and gravels.	 --- ? --- 'Gravels,' ? ?	Marine clays; poorly exposed. --- Interbedded silt, sand, and gravel with peat beds. Vegetation--- forest-tundra and tundra.	Marine clays --- Interbedded clay, silt, sand, gravel, and peat. Vegetation--- boreal forest, forest-tundra, and shrub tundra.	Freshwater silts overlain and underlain by marine sediments; ice-wedge casts. Vegetation--- forest nearby(?). ---

Fig. 2–9. Stratigraphic correlation of pre-Buckland Pleistocene sediments along the Yukon Coastal Plain between Herschel Island and Sabine Point (from Rampton 1982, Tab. 14, Erratum page). Published with permission of the Minister of Supply and Service Canada.

Herschel Island, Yukon, Canada

Herschel Island is located at the western end of Mackenzie Bay of the Beaufort Sea, northern Yukon, Canada (fig. 2–8). It is part of the Yukon Coastal Plain, a generally low, nearly flat area mostly < 60 m in elevation. Herschel Island is most striking; it reaches a maximum elevation of 181 m (596 feet), covers > 100 km^2, and is bounded by steep sea cliffs > 60 m high on its northern side. It is composed almost entirely of Pleistocene sediments derived from Herschel Basin and deformed by ice pushing from the southeast (Mackay 1959; Rampton 1982).

Herschel Island plus Herschel Basin represent one of the largest hill-hole pairs in the world. A long ridge of similar ice-shoved Pleistocene sediments extends on the mainland from Kay Point to King Point reaching maximum elevations locally > 75 m. The Yukon Coastal Plain is covered by tundra and underlain by continuous permafrost with a thickness generally > 300 m (Rampton 1982). The deformed Pleistocene sediments of Herschel Island and the Kay-King Points ridge are consolidated by this permafrost.

The ice-pushed structures of Herschel Island and the Kay-King Points ridge are composed largely of preglacial Pleistocene sediments. No pre-Quaternary bedrock is exposed in any of these areas. The deformed strata consist of interbedded clay, silt, sand, gravel, and peat that were deposited in shallow marine, lacustrine, delta, lagoon, or flood-plain environments prior to the Buckland Glaciation. Regional correlation of these strata is hampered by discontinuous exposures, frequent facies changes, and glaciotectonic deformations. Nonetheless, a general stratigraphic framework is now available (fig. 2–9).

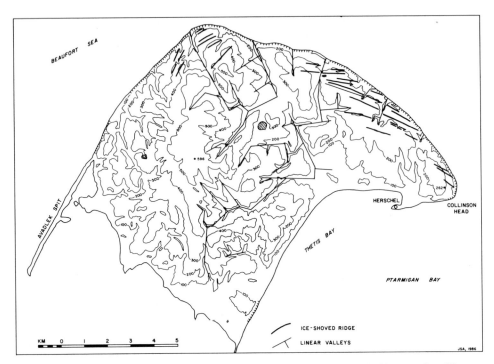

Fig. 2–10. Topographic map of Herschel Island showing ice-shoved ridges and trellis drainage pattern. Elevations in feet; contour interval = 100 feet (app. 30 m); diagonal lining shows larger lakes. Based on interpretation of aerial photographs.

Bouchard (1974) and Rampton (1982) divided the pre-Buckland strata on Herschel Island into three units: (1) upper unit consisting mostly of marine sediments, (2) a middle unit of mixed sediments, and (3) a lower marine-clay unit. Although variable in thickness, the three units may comprise up to 50 m of strata.

The Buckland Glaciation is represented by pebbly till, kame deltas and terraces, various moraines, melt-water channels, and other glacial features, all subdued in morphologic expression. The Buckland ice limit descends northwestward from the Richardson Mountains, to the Buckland Hills, to just beyond the Firth River delta west of Herschel Island (fig. 2–8). Glacial erratics near the crest of Herschel Island show that the island was covered during the maximum Buckland advance.

The Buckland Glaciation was presumed to be early Wisconsin in age by Rampton (1982), although Dyke and Prest (1987b) indicated a late Wisconsin age of 25,000 years BP. The upper dislocated marine unit on Herschel Island was deposited during a somewhat warmer, high sea-stand before the Buckland Glaciation. This corresponds to the Pelukian transgression of Alaska, thought to be Sangamon and early Wisconsin in age. The middle and lower sedimentary units were deposited during still earlier glaciations and interglaciations.

Herschel Island is overall an arcuate, asymmetrical dome, concave toward the southeast with the crest offset west of center (fig. 2–10). The northeastern and

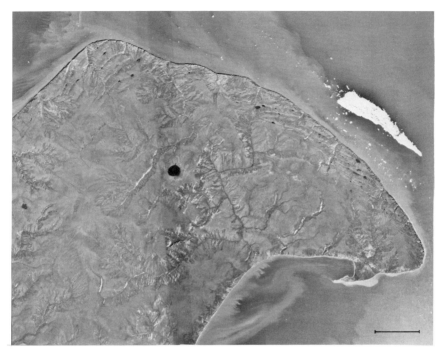

Fig. 2–11. Aerial photograph of northern and eastern Herschel Island, northern Yukon. Note ice-shoved ridges along northern margin and distinctive trellis drainage pattern. Scale bar is 1.5 km long; north toward top. Aerial photographs A24123–140 and 149; copyright 1975. Her Majesty the Queen in Right of Canada, reproduced from the collection of the National Air Photo Library with permission of Energy, Mines and Resources Canada.

north-central margins of the island display conspicuous ridges, which form an arcuate pattern, concave southward, following the northern edge of the island (fig. 2–11). The pattern of ridges is emphasized by the presence of very small, elongated lakes within intervening valleys. Elsewhere, the land slopes fairly uniformly toward the coast.

Deep stream valleys and ravines form a distinctive trellis drainage pattern in the northern and central portions of the island east of the crest (Mackay 1959). Smaller ravines parallel ice-shoved ridges, but larger valleys crosscut the ridges at nearly right angles. The larger ravines display a radial pattern with an apex near the concave side of the island on Thetis Bay. The valleys and ravines are the results of post-Buckland erosion of softer strata and fracture zones. Thus, the present landscape of Herschel Island is in part an erosional morphology adjusted to underlying structures.

The most common structures revealed in sea cliffs are low-angle thrust faults and open synclines and anticlines (Mackay 1959). Tilted beds usually show apparent dips of 5–20°, but are nearly vertical in a few places. Overturned folds, repetition of beds, and inverted strata are also present in some locations. Shear planes with slickensides are especially common in clay beds. The available observations demonstrate a close agreement between internal structure and external morphology of Herschel Island. The island represents a single episode of ice pushing by an ice lobe moving from the southeast.

The structures of Herschel Island include segregated ice layers and sheets, up to several m thick, which parallel the bedding of deformed strata. Crystallographic evidence proves that these ice layers developed in horizontal positions and were subsequently deformed along with the enclosing sediments by glacier pushing (Mackay and Stager 1966; Mackay et al. 1972). The ground ice probably formed when permafrost developed immediately before the Buckland Glaciation (Rampton 1982), and permafrost has persisted until the present.

Based on elevation of the Buckland ice limit, Rampton (1982) constructed paleocontours on the maximum Buckland ice surface (fig. 2–8). The 500-foot contour lies just beyond the western edge of Herschel Island, so maximum ice elevation at Herschel Island is estimated to have been 600 feet. This happens to coincide with the highest elevation on Herschel Island itself, 596 feet. Ice thickness over the island may have been some greater, however, due to crustal depression beneath the ice lobe.

The pre-Buckland sediments were undoubtedly permafrozen at the time of ice shoving. However, the age and thickness of the permafrost is uncertain. Climatic indicators in the pre-Buckland sediments suggest that conditions slightly warmer than today existed at certain times prior to Buckland Glaciation. Mackay (1959) and Rampton(1982) both agreed that ice loading caused compression and increase in hydrostatic pressure in unfrozen sediments below the permafrost.

Thrust blocks may have been detached at the base of the permafrost and pushed over a cushion of high-pressure fluid. This implies that permafrost at the time of thrusting was only 50-60 m deep, the maximum thickness of disturbed sediment. Mathews and Mackay (1960) pointed out, however, that clay may remain plastic at sub-zero temperatures. Thus, thrusting could have taken place along clay layers within still thicker permafrost.

Eyjabakkajökull, Iceland

The ice-marginal area of Eyjabakkajökull, Iceland, is an ideal site for studying glaciotectonic phenomena. Three types of landforms are found within a restricted area: cupola-hills, hill-hole pairs, and small composite-ridges (fig. 2–12). The most outstanding and intensively studied features are the small composite-ridges (Chapter 4). They also represent a form of hill-hole association, for which a direct genetic link can be established due to their recent age (< 100 years old). However, as will be shown, there are discrepancies between the volumes of composite-ridges and their associated holes or basins.

Fig. 2–12. Aerial photograph of the snout vicinity of Eyjabakkajökull, a northeastern outlet glacier of Vatnajökull. Scale bar is 500 m long; north shown by arrow. Airphoto copyright Landmaelingar Islands 1967; reproduced with permission.

Eyjabakkajökull is an outlet glacier of Vatnajökull Ice Cap in southeastern Iceland (Map 2). Although of moderate size by present-day standards (10 km long by 2.5 km wide), Eyjabakkajökull is small when compared with Pleistocene ice sheets and glaciers. Consequently, glaciotectonic features at this site are similarly small in size. The position of the ice margin has fluctuated over a distance of several km in the last century as a consequence of the glacier's surge behavior (Paterson 1981). Surges have occurred at regular intervals (1890, 1931, 1972), but later advances have been progressively smaller since the 1890 maximum.

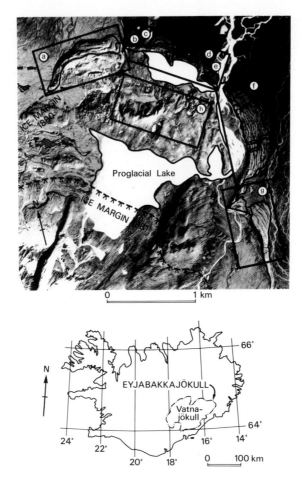

Fig. 2–13. Snout area of Eyjabakkajökull showing subdivision of ice-shoved ridges into individual sets; taken from Croot (1988b, fig. 1).

Eyjabakkajökull discharges into a broad flat valley floor that is underlain by a range of sandur sediments. With the exception of recent (post-1890) ice-proximal deposits, these sediments are mostly fine-grained sand, silt and clay, in which texture varies over short vertical and lateral distances. Interbedded tephra layers serve as markers for correlating disturbed and undisturbed sections.

Seven sets of small composite-ridges were formed as the glacier reached its maximum in 1890. These composite-ridges are composed of dislocated sandur sediments and describe a broad arc across the valley at and immediately beyond the 1890 ice margin (a-g, fig. 2–13). The dimensions, and thus volume, of each ridge set vary, but an upvalley/upglacier depression is associated with each set. Some of these depressions are now partly filled with sediments that postdate the formation of the composite-ridges; others contain small, shallow lakes.

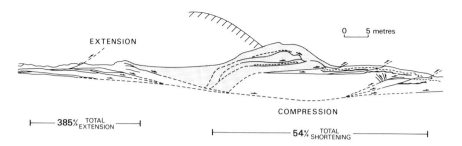

Fig. 2–14. Section through ridge set e (fig. 2–13) showing that portion (shaded) of the ridge transported more than 10 m from its original (predeformation) site.

Each ridge set is visually proportional in area to the basin that lies upglacier from it. However, on closer inspection it becomes clear that only a portion of the material comprising the composite-ridges has been transported from the basins. A simplified section, parallel to ice movement through one of the ridge sets, reveals that only part of the material (shaded portion, fig. 2–14) was transported any distance from its source. The remainder was thrust forward proglacially, and the transport distances for these thrust sheets are < 5 m each.

The material removed from the upglacier basin is more deformed and represents about 50% of the cross-sectional area (shaded portion, fig. 2–14). Assuming this is a roughly consistent proportion, then 50% of the ridge volume was derived from the basin, amounting to 50,000 m³ (fig. 2–15, Tab. 2–1). The basin immediately upglacier from ridge set e has a surface area of approximately 25,000 m², which corresponds to removal of 2 m of material from the area of the lake basin. This

Table 2–1. Morphometric data for ridge sets (a–g, fig. 2–13) from 1890 advance of Eyjabakkajökull, Iceland. Adapted from Croot (1987, Tab. 1).

Ridge Set	H (m)	L (m)	W (m)	$\theta°$	Aspect Ratio	Volume (10^5 m³)	Mass (10^8 kg)
a	40	220	360	7°	5.5	15.8	26.9–37.9
b	7	50	111	9°	7.1	0.2	0.3–0.4
c	15	230	330	3.5°	15.3	5.7	9.7–13.6
d	8	200	305	4°	25	2.4	4.2–5.9
e	9	110	200	8°	12.2	1.0	1.7–2.4
f	5	280	527	2°	56	3.7	6.3–8.8
g	5	270	270	2°	54	1.8	3.1–4.4

Note: Aspect ratio = L/H. Volume estimated to be V = LWH/2 in m³. Density values probably range from 1.7 for silt to 2.4 for gravel; min-max range is given for mass.

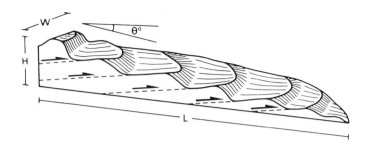

Fig. 2–15. Schematic diagram showing morphometric factors used for calculating volume of ridge sets (Table 2–1). Taken from Croot (1987, fig. 5).

value is consistent not only with depth of the lake basin, but also with the thickness of individual thrust floes, which confirms a uniform depth to a plane of decollement.

The same principles of calculation can be successfully applied to the other composite-ridges around the 1890 margin. It is interesting to note that the cupola-hill (h, fig. 2–13) situated behind the 1890 margin is also associated with a large, deep lake. This hill appears to have once been part of a prominent composite-ridge system marking an older ice-margin position. This hill-hole pair was overridden and smoothed into a cupola-hill by the 1890 advance. It is evident from these examples in Iceland that hill-hole pairs may have complex internal structures and histories of development.

CHAPTER 3

LARGE COMPOSITE-RIDGES

Introduction

The most typical and distinctive glaciotectonic landforms are ice-shoved ridges found in many glaciated plains. Prest (1983:45) aptly described such ridges as, 'a composite of great slices of up-thrust and commonly contorted sedimentary bedrock that is generally interlayered with and overlain by much glacial drift.' The term *composite-ridges* (= transverse-ridges of Clayton *et al.* 1980) is used here for such ice-shoved ridges. Composite-ridges that include a substantial volume of deformed pre-Quaternary bedrock should not be called end moraines.

Composite-ridges are here divided into large (> 100 m relief) and small (< 100 m relief) categories. Large composite-ridges may be up to 200 m high, 5 km wide and 50 km long (fig. 3–1). In map view, composite ridges are often arcuate and concave upglacier with a radius of curvature of 2 to 10 km (Clayton *et al.* 1980). Individual ridges typically display several 10s of m of topographic relief and are a few 100 m in width. Ridges and intervening valleys, which often contain small elongated lakes, form a subparallel pattern that follows the general curved outline of the ice-shoved hill. This arcuate pattern marks the margin of the ice lobe or tongue that shoved up the ridges.

The ridges are developed on the crests of folds or the upturned ends of thrust blocks. A close correspondence typically exists between structural features and

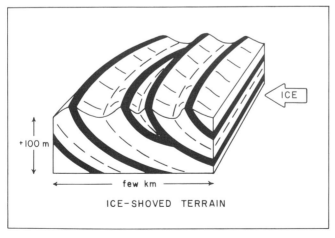

Fig. 3–1. Schematic block diagram of typical thrust structure of large composite-ridges. Arcuate pattern reflects movement of an ice lobe from the right.

Fig. 3–2. Aerial photograph of Flade Klit, northern Denmark. Composite-ridges form an open crescent, concave northward. Location of cliff exposure at Hanklit (Plate I) shown by arrow; Salgjerhøj (solid dot) is high point at 88 m; adjacent water body is the Limfjord estuary. Scale bar is 500 m long; north toward top. Copyright Geodætisk Institut, Denmark, 1986 (A. 29/88).

topography. Large composite-ridges usually involve considerable disruption of pre-Quaternary bedrock, which may comprise a major volume of the ridges. However, the depth of structural disturbance is generally not greater than about 200 m (Kupsch 1962).

The folds and thrust blocks that form ridges have usually been detached, transported some distance, and stacked up in an imbricated structure. Composite-ridges are, thus, allochthonous in a glaciotectonic sense, and it may be possible to recognize the upglacier depression from whence material in the ridges was derived. However, in many cases a discrete source basin cannot be specifically identified, as with the hill-hole pair.

The typical morphology and structure of large composite-ridges is displayed at Flade Klit on the island of Mors, northwestern Denmark (fig. 3–2). Paleocene bedrock and drift were folded and thrust into composite-ridges during late Weichselian glaciation (Gry 1940). The bedrock, consisting of clayey diatomite inter-

Fig. 3-3. Photograph of southern Møns Klint from the beach showing chalk masses of Sommerspiret (center) and Nælderendenakke (left) standing > 100 m high. Photo by J. Aber, 1986.

bedded with volcanic ash layers, was especially susceptible to ice-push deformation.

Large, overturned, rootless folds of bedrock and drift were thrust up forming ridges, as at Hanklit (Plate I). Maximum elevation at Salgjerhøj (88 m) is 100 m above the floor of the Limfjord estuary immediately to the north. The estuary basin is presumably the source of material now in Flade Klit, although a specific source depression has not been identified. The composite-ridges form a gentle arc, concave toward the north.

Large composite-ridges are topographically and structurally similar to such thrust and folded mountain belts as the Canadian Rockies or Swiss Juras that were formed by thin-skinned tectonics. The only real difference is size, ice-shoved ridges being one or two orders of magnitude smaller than true mountains. The following examples of large composite-ridges are, in fact, miniature mountains produced by ice shoving of soft sedimentary bedrock.

Møns Klint, southeastern Denmark is undoubtedly the most famous and spectacular of all glaciotectonic sites. Large scales of chalk and drift are beautifully

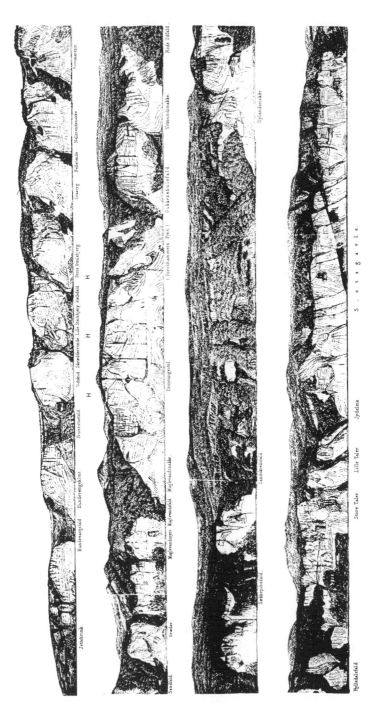

Fig. 3–4. Møns Klint section as viewed from the east; Jættebrink at southern end, Slotsgavle at northern end. Black lines within blank chalk masses show deformed flint layers. Reproduction of copper engraving by Puggaard (1851).

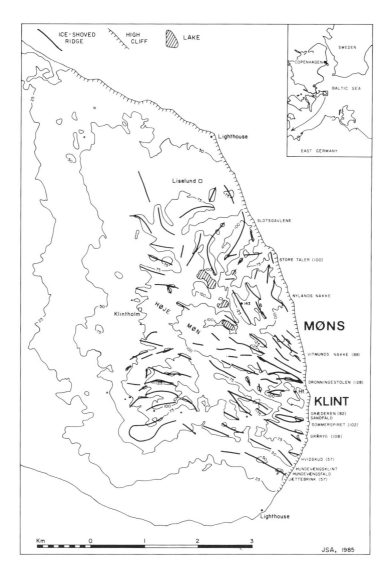

Fig. 3-5. Topographic map of eastern Møn showing ice-shoved ridges. Individual chalk cliffs and inter-chalk falls are indicated with heights of chalk cliffs given in parentheses. Contour interval = 25 m; Ht. = Hotel. Taken from Aber (1985, fig. 1).

exposed in a scenic cliff (fig. 3-3) and form a rugged landscape inland from the cliff. The combination of cliff exposures and composite-ridge morphology provides a 3-dimensional display of Møns Klint's glaciotectonic structure.

Prophets Mountains, central North Dakota are very similar in size and topographic expression to the composite-ridges at Møns Klint. However, the internal structure of Prophets Mountains is not well known. The Dirt Hills and

Fig. 3–6. Photograph of central Møns Klint from the beach showing chalk mass of Dronningestolen. Photo by J. Aber, 1986.

Cactus Hills, southern Saskatchewan, are still larger examples of composite-ridges built mainly of upper Cretaceous bedrock. All of these examples are Pleistocene in age; similar large composite-ridges have not been created by relatively small Holocene glacier advances.

Møns Klint, Denmark

The large chalk cliffs of eastern Møn, southeastern Denmark are justifiably famous for their scenic beauty and distinctive geological structure. The cliffs are perhaps the finest example of glaciotectonic features in the world, and they have a long history of geological study beginning with Agricola (1546). The earliest known illustration appeared in the 1700s (fig. 1-2). Johnstrup (1874) first recognized the glaciotectonic genesis of Møns Klint, and the first map showing ice-shoved ridges on Møn was published by Haarsted (1956). The copper engraving of Møns Klint by Puggaard (1851) remains a classic illustration of the cliff (fig. 3–4).

The high, rugged terrain, known as Høje Møn, located at the eastern end of the island is composed of several dozen chalk scales that were piled up during late Weichselian ice advances 20,000 to 13,000 years BP (Sjørring 1981). Høje Møn generally exceeds 75 m elevation reaching a high at 143 m (fig. 3-5). Dronningestolen (the Queen's throne), the largest chalk cliff in the center of Møns Klint, is 128 m high (fig. 3-6). Presumably undisturbed chalk bedrock was intersected in drill holes at -20 to -40 m (Haarsted 1956), so > 150 m of structural relief is indicated.

The individual scales exposed in Møns Klint consist of upper Cretaceous (Maastrichtian) white 'writing chalk' that was deformed along with drift. Chalk now forms ridges and cliffs because of its greater resistance to erosion, whereas intervening drift has been eroded into valleys that form the falls along the coast (fig. 3-7). The chalk is very uniform in lithology, aside from occasional layers of flint nodules, and thus, stratigraphic correlation between chalk masses is difficult at best. Surlyk (1971) divided the Danish Maastrichtian into ten brachiopod biozones: zones 1-7 are lower Maastrichtian and 8-10 are upper Maastrichtian. These biozones are the best method for establishing correlation between chalk masses.

The drift strata are well exposed in only a few places, notably at Hundevængsfald and Sandfald. The composite sequence includes from the top down (Hintze 1937; Surlyk 1971): (1) discordant drift, (2) upper dislocated till, (3) stone-poor clay, (4) cross-bedded sand, and (5) lower dislocated till. Konradi (1973) concluded that the lower dislocated till and all the overlying drift must be Weichselian in age, because the lower till contains abundant reworked Eemian foraminifera.

The stone counts carried out by Hintze (1937) show that the lower dislocated till had a Baltic source, whereas the upper dislocated till was deposited from the northeast by an ice advance which caused considerable local erosion. Both of these tills were deposited before the southern portion of Møns Klint was thrust.

Møns Klint and Høje Møn can be divided into three morphostructural regions on the basis of ridge morphology, cliff structures, and chalk stratigraphy (Aber 1988a). The southern region includes Jættebrink through Sommerspiret, with chalk biozones 3, 4 and 5. The chalk scales form a series of imbricately thrust anticlines that dip southward and are increasingly deformed toward the north. These chalk masses continue inland as long, straight to arcuate ridges, concave toward the south. This region was thrust by ice movement directly from the south.

Dronningestolen along with Græderen and Maglevandspynten make up the central portion of the cliff. Dronningestolen is a huge composite of many lesser chalk floes (biozones 3 through 8) folded and stacked on top of each other in the overall form of a broad anticline. Dronningestolen continues inland as a massive ridge, beyond which the central region is marked by many short, offset ridges. The central region was apparently deformed by ice pushing from both north and south.

The northern region of Vitmunds Nakke through Slotsgavlene includes biozones 5 through 8 in scales oriented oblique to the coast. A regular shift is displayed in

Fig. 3–7. Aerial photograph of Møns Klint and Høje Møn, southeastern Denmark. Chalk masses form high cliffs and continue inland as sharp-crested ridges. More rugged portion of ridges covered by beech forest; presence of chalk masses in fields shown by light tone. Scale bar is 500 m long; north toward top. Copyright Geodætisk Institut, Denmark, 1974 (A. 29/88).

structural strike along the cliff from southeast in the south, to south in the center, to southwest in the north. This corresponds to the arcuate pattern of ridges inland from the cliff. Thrusting of the northern region was evidently brought about by ice advance from the east or east-northeast.

The late Weichselian glaciation of southern Denmark occurred in distinct phases separated by brief ice retreats (Berthelsen 1978; Houmark-Nielsen 1987): (1) Old Baltic advance from the southeast that deposited the Ristinge Klint Till, but caused little glaciotectonic disturbance; (2) Main Weichselian phase with initial advance

Fig. 3–8. Ice-shoved hills of southern Saskatchewan and adjacent Alberta, Canada. 1 = Radville area, 2 = Dirt Hills, 3 = Cactus Hills, 4 = Chaplin Lake area, 5 = The Coteau, 6 = Lancer area, 7 = Monitor area, 8 = Neutral Hills, 9 = Killarney Lake area. Adapted from Kupsch (1962, fig. 1).

from the northeast (Mid Danish Till) and Storebælt readvance from the east (North Sjælland Till), both of which resulted in major glaciotectonic disruptions; (3) Young Baltic advances coming first from the southeast (East Jylland Till) with late readvances by ice tongues along major straits from the south (Bælthav Till), both of which created much ice-push deformation.

The upper and lower dislocated tills at Møns Klint are correlated respectively with the Mid Danish and Ristinge Klint Tills. The overlying discordant drift probably relates to the Young Baltic advances. Thrusting of the southern region of Høje Møn was the final glaciotectonic disturbance and must be associated with a late Young Baltic (Bælthav) advance coming from the south. Earlier thrusting of the northern region is related to either the Main Weichselian phase, most likely the Storebælt readvance from the east, or the East-Jylland phase of the Young Baltic. In both cases, ice advances overrode Møn following the thrusting of composite-ridges. Postglacial erosion has since etched out the softer Quaternary strata and produced the present distinct morphology of Høje Møn.

Dirt Hills and Cactus Hills, Saskatchewan, Canada

Large composite-ridges of the Møns Klint type are well developed in southern Saskatchewan and adjacent Alberta on the Canadian Plains. The greatest development of composite-ridges is found in the Dirt Hills and the Cactus Hills (areas 2 and 3, fig. 3–8). The Dirt Hills reach a maximum elevation of 880 m (2887 feet), some 300 m above the Regina Lake Plain to the north and 150 m above the adjacent Missouri Coteau. The Cactus Hills are nearly as high in elevation. The Dirt Hills and Cactus Hills together encompass a region approximately 1000 km^2 in extent.

The Missouri Coteau is a major northeast-facing escarpment that marks the boundary between the Saskatchewan and Alberta Plains. The Missouri Coteau as well as lower hills to the northeast are bedrock features resulting from preglacial erosion and subsequently modified by glaciation. A mantle of drift covers most of the region today.

Higher elevation of the Alberta Plain is a reflection of more resistant terrestrial sandstone interbedded with mudstone and lignite. The Saskatchewan Plain, in contrast, is underlain by softer marine shale. The regional geologic structure consists of essentially flat-lying strata which dip very gently to the east or northeast (Fraser *et al.* 1935). Steeply dipping, folded, and faulted bedrock structures are common, however, in ice-shoved hills along the Missouri Coteau.

The disturbed bedrock structures of the Dirt Hills were apparently first noted by Bell (1874), and Fraser *et al.* (1935) described faulted bedrock in the northern Dirt Hills near Claybank. The true glaciotectonic origin of the Dirt Hills and Cactus Hills was first demonstrated by Byers (1959), who believed the deformations had been produced by subglacial frictional drag.

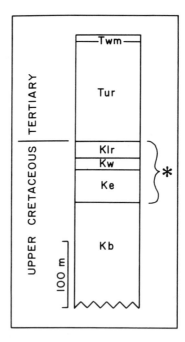

Fig. 3–9. Stratigraphic column for bedrock of the Dirt Hills and Cactus Hills vicinity. Asterisk indicates deformed units. Based on Fraser *et al.* (1935) and Parizek (1964).

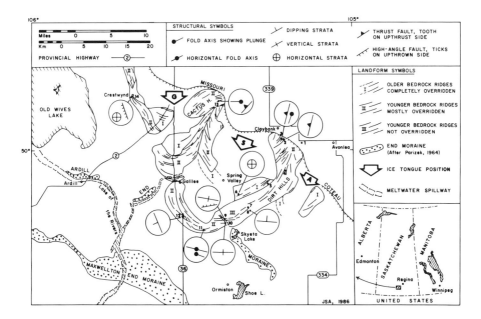

Fig. 3–10. Map of Dirt Hills and Cactus Hills vicinity showing pattern of ice-shoved ridges, locations of sites, structural features, and ice tongues: G = Galilee, S = Spring Valley, A = Avonlea. Modified from Aber (1988c, fig. 3).

A general analysis of the genesis of these hills along with similar ice-shoved hills in western Canada was presented a short time later by Kupsch (1962). Christiansen (1961) and Parizek (1964) described the ice-pushed ridges in conjunction with regional resource mapping, and the great depth of structural disturbance was confirmed by test drilling on the crest of the Dirt Hills (Christiansen 1971a; Christiansen and Whitaker 1976).

Three upper Cretaceous bedrock units can be seen deformed in the Dirt Hills and Cactus Hills. These three in ascending order are (fig. 3–9): Eastend (Ke), Whitemud (Kw), and lower Ravenscrag (Klr). The deformed strata are predominately terrestrial, bentonitic or kaolinitic sandstone, plus mudstone and lignite, having a total thickness up to 90 m. They are underlain by presumably undisturbed marine shale of the Bearpaw Formation (Kb).

Folded and thrust bedrock scales stacked in an imbricated pattern comprise the overall structure of the Dirt Hills and Cactus Hills (fig. 3–10). At most sites, drift covers the bedrock but is not involved in the deformations. The drift is dominated by dolostone and crystalline erratics in the pebble fraction. However, at two sites (6 and 9) older blocks of drift rich in quartzite and chert are exposed as a result of glaciotectonic uplift. Not only have scales been displaced horizontally, but considerable vertical movement has also occurred. Maximum structural uplift is documented at site 9, where a block of Eastend Formation is standing vertically some 200 m above its normal stratigraphic position (Aber 1988c).

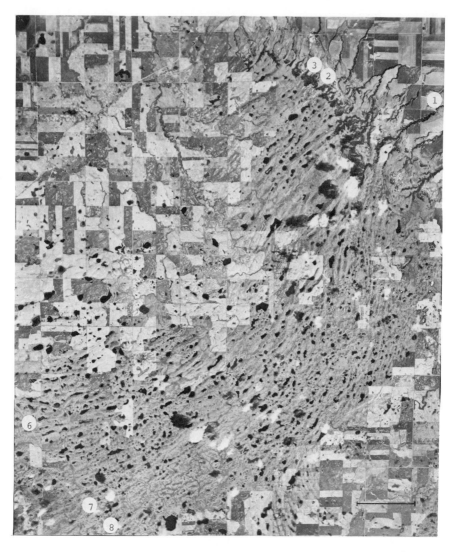

Fig. 3–11. Aerial photograph of eastern and northern Dirt Hills, Saskatchewan. Numbered sites same as Fig. 3–10. Very small elongated lakes occupy narrow valleys between ice-shoved ridges. Scale bar is 2 km long; north toward top. Aerial photo A21657–20; copyright 1970. Her Majesty the Queen in Right of Canada, reproduced from the collection of the National Air Photo Library with permission of Energy, Mines and Resources Canada.

Remarkable agreement exists between orientations of bedrock structures, trends of individual ridges, and overall outlines of the Dirt Hills and Cactus Hills. This amply confirms Kupsch's (1962) conclusion that ice-shoved ridges on the Missouri Coteau are direct or first-order morphologic expressions of bedrock structures produced by ice pushing.

Bedrock structures at most sites are related to a single episode and direction of ice pushing, but some sites (13) show evidence for multiple phases of deformation. Variations in bedrock competence clearly influenced structural development. Thrust faults are usually located within lignite or claystone beds; conversely, thicker sandstone layers comprise the larger folds and fault blocks.

Given the fact that ridges are first-order morphologic expressions of bedrock structures, the primary regional structure of the Dirt Hills and Cactus Hills can be directly determined from topography. Ridge morphology is particularly evident on aerial photographs (fig. 3–11) due to the relatively dry climate and sparse vegetation. Because of their great size, the Dirt Hills and Cactus Hills are also conspicuous on satellite images (Plate II).

Three classes of ice-shoved ridges are identified on the basis of topographic expression and structural information. The three ridge classes include a subdued class I and two prominent classes, II and III (fig. 3–10). Ridges of classes I and II were overridden and smoothed by ice, whereas the highest class III ridges of the southern Dirt Hills were never overrun. Class I ridges were presumably created during an earlier glacier advance of unknown age. Class II and III ridges were thrust up by the last ice advance to push onto the Missouri Coteau.

The Dirt Hills and Cactus Hills form two large loop-shaped ranges which along with nearby hills define the outlines of three ice tongues that caused the thrusting of bedrock ridges. These three ice tongues were: (1) Galilee, west of Cactus Hills, (2) Spring Valley, between Cactus Hills and Dirt Hills, and (3) Avonlea, east of Dirt Hills.

Thrusting of the Dirt Hills and Cactus Hills did not happen during initial advance of the late Wisconsin Lostwood ice sheet, which reached its maximum position near the United States border about 17,000 years BP (Christiansen 1979; Dyke and Prest 1987b). At that time, the Dirt Hills and Cactus Hills did not yet exist. Thrusting of the hills occurred later, probably around 13,000 years BP (Fenton, Moran *et al.* 1983; Dyke and Prest 1987b), during a strong readvance of the Weyburn lobe (Christiansen 1956). This lobe generated lateral ice tongues that pushed into embayments of the Coteau. Thrusting of bedrock occurred around the margins of these ice tongues due to rapid loading and forward movement.

All three ice tongues caused thrusting of class I bedrock ridges during an earlier advance, however the time and nature of this advance are uncertain. The main thrusting of class II and III ridges occurred during a readvance of the Galilee and Spring Valley ice tongues. These ice tongues overran class I ridges and thrust up new ridges to the south forming higher portions of the Dirt Hills and Cactus Hills. The Avonlea ice tongue also readvanced at this time, but without thrusting up any new ridges.

The Galilee and Avonlea ice tongues reached positions marked by the Ardill end moraine and Lake of the Rivers spillway during the readvance. The Spring Valley ice tongue, however, stopped on the inner (northern) side of the Dirt Hills, from where a spillway was cut across class III ridges toward Skyeta Lake (fig. 3–12).

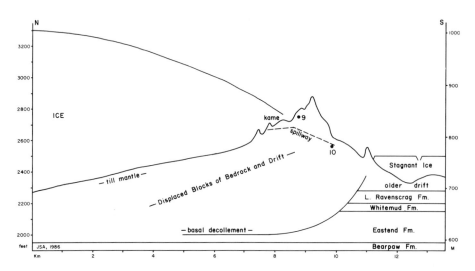

Fig. 3–12. Schematic model for thrusting of southern Dirt Hills. North-south profile of present land surface through high ridges west of Skyeta Lake spillway (shown by dashed line; numbered sites same as fig. 3–10). Basal decollement is located in claystone beds of the lower Eastend Formation. Profile of ice tongue based on assumed minimum thickness of 300 m at northern end of section. From Aber (1988c, fig. 12).

The class III ridges, thus, formed a nunatak between active ice to the north and older stagnant ice lying on the Coteau to the south.

Building of the Ardill end-moraine system and cutting of associated spillways were related to the same ice advances that caused the main phase of thrusting in the Dirt Hills and Cactus Hills. Following thrusting of the hills, building of the end moraine, and cutting of the spillways, the ice tongues stagnated and downwasted leaving an irregular accumulation of hummocky moraine over much of the area north of the ice margins.

Rapid loading of competent sandstone bedrock (lower Ravenscrag, Whitemud, and upper Eastend Formations) over saturated, incompetent mudstone strata (lower and middle Eastend Formation) caused thrusting around the margins of the ice tongues. The fact that bedrock was most likely thawed and saturated is confirmed by abundant evidence for melt-water spillways, large proglacial lakes, and wasting stagnant ice masses throughout southern Saskatchewan during deglaciation. There is no evidence that permafrost existed during the late Wisconsin deglaciation. Ubiquitous melt-water means that thrust blocks could not be moved by freezing onto the undersides of the ice tongues, but were displaced by squeezing out from under the ice margin.

Prophets Mountains, North Dakota, United States

Prophets Mountains are a premier example of large composite-ridges, located in

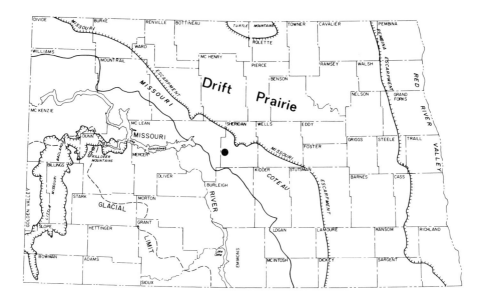

Fig. 3-13. Map of North Dakota showing major physiographic features and counties. Glacial limit shows outermost pre-Wisconsin glaciation; Wisconsin drift is mainly found north and east of Missouri River. Location of Prophets Mountains shown by solid dot. Adapted from Clayton *et al.* (1980).

southwestern Sheridan County, North Dakota (fig. 3-13). This area is part of the Missouri Coteau, a hilly upland belt northeast of the Missouri River, which extends northwestward into Saskatchewan. The Drift Prairie northeast of the Missouri Coteau generally lies about 500 m in elevation, whereas elevations on the Coteau are mostly in the vicinity of 600 m. Prophets Mountains reach a maximum elevation of 690 m (2250 feet) and cover roughly 20 km^2 (fig. 3-14). The history of geologic study for Sheridan County was reviewed by Bluemle (1981), who described the morphology and structure of Prophets Mountains.

Sub-drift bedrock in Sheridan County consists of upper Cretaceous Hell Creek Formation and lower Tertiary Cannonball Formation (fig. 3-15). The Hell Creek Formation is comprised of interbedded sandstone, mudstone, carbonaceous shale, and thin lignite of terrestrial origin. The overlying Cannonball Formation includes marine carbonaceous and lignitic siltstone, shale, and micaeous sandstone. The contact between the two normally flat-lying formations is intersected in test holes near Prophets Mountains at about 505 m elevation.

The Coleharbor Group in North Dakota (Clayton *et al.* 1980) includes all Pleistocene glacial and glacially derived sediments. In Sheridan County, the Coleharbor Group ranges from < 30 m to > 200 m in thickness and consists primarily of till. Till texture varies considerably within the county, but it has proven impossible to identify individual till sheets on the basis of pebble composition or texture. Till is the predominate glacial sediment over much of the Prophets

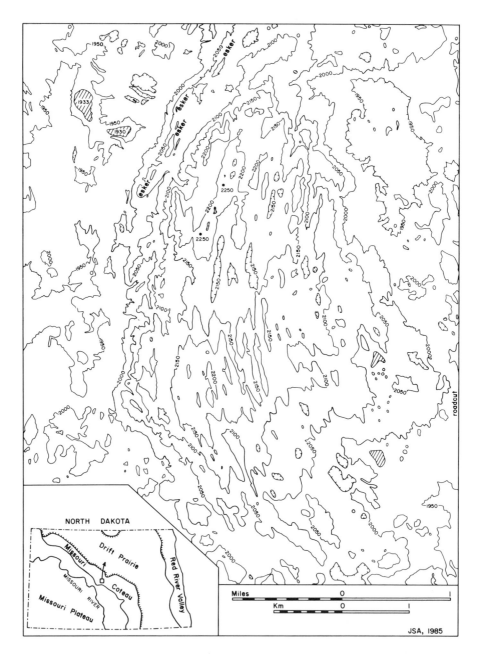

Fig. 3–14. Topographic map of Prophets Mountains vicinity. Multiple, parallel, north-south ridges comprise the central portion of Prophets Mountains. Contours in feet; contour interval = 50 feet (app. 15 m). Perennial lakes shown by diagonal lining; deformed bedrock exposed at road cut on eastern edge.

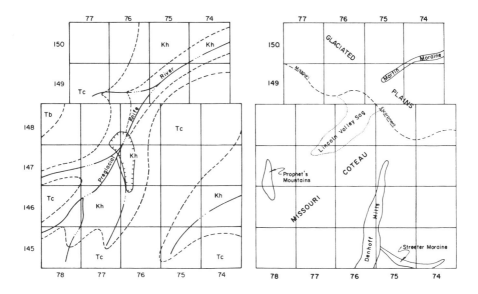

Fig. 3–15. Maps of Sheridan County: right – physiographic features, left – generalized bedrock topography (dashed lines) and possible preglacial drainage. Kh = Hell Creek Formation, Tc = Cannonball Formation; closed bedrock depression shown by tick-marked line. Marginal numbers indicate townships and ranges; each township is 6 miles (app. 10 km) on a side. Adapted from Bluemle (1981, figs. 1 and 20).

Mountains vicinity. Sand and gravel sediments cover about 15% of the county and amount to a similar fraction of total thickness of the Colcharbor Group. Along the northwestern flank of Prophets Mountains, a series of esker ridges is preserved (fig. 3–14).

Prophets Mountains consist of multiple, parallel ridges trending generally north-south with a slight arcuate tendency, concave toward the east. The base of Prophets Mountains is roughly outlined by the 2000-foot contour, with higher elevations and steeper slopes toward the western side of the hills. Total relief exceeds 100 m, and individual ridges are typically 10-25 m above adjacent valleys.

Folded and contorted bedrock was previously exposed in a road cut on the southeastern edge of Prophets Mountains (Bluemle 1981). About 15 m of Hell Creek and 12 m of Cannonball strata are present at an elevation approximately 100 m above the normal contact of these two formations. The crest of Prophets Mountains is 80 m still higher, so considerable structural uplift by ice pushing seems probable.

The simple arcuate pattern of ridges within Prophets Mountains corresponds to ice pushing by a single ice tongue advancing directly from the east. An adjacent source depression is not visible at the surface; however, a likely source depression is buried in the western portion of township 147, range 76 (fig. 3–15). Test drilling has revealed a closed bedrock depression of about 75 km^2 as much as 50 m below the presumed preglacial valley floor.

Unusually thick drift, up to 220 m deep, fills this depression. The depression is located roughly 18 km directly east of Prophets Mountains. Material from this depression could have supplied floes for building ice-shoved ridges to the west at Prophets Mountains and to the southeast in Denhoff Hills (Bluemle 1981).

Prophets Mountains is located above a large, complex aquifer developed partly in Cretaceous sandstone and partly in a buried valley tributary to the preglacial Knife River (fig. 3–15). At present, ground water drains toward the east through this aquifer, but eastward drainage was blocked by glacier ice during thrusting of Prophets Mountains. Ground-water flow was reversed, and pore-water pressure built up enough to facilitate glacier thrusting, according to Bluemle and Clayton (1984). The scarcity of exposures and inability to discriminate different till sheets renders any interpretation of the genesis of Prophets Mountains somewhat speculative, however. The distinct morphology of Prophets Mountains suggests a young, probably Wisconsin, age.

CHAPTER 4

SMALL COMPOSITE-RIDGES

Introduction

Small composite-ridges are perhaps the most common glaciotectonic landform. They display the same morphologic traits and structural features as do large composite-ridges. Small composite-ridges are also found in similar topographic settings, such as escarpments, islands, or valley sides. A source depression is located a short distance upglacier in some cases. Finally, both large and small composite-ridges are usually associated with ice margins marking glacier stillstands or readvances.

The size division between large (> 100 m relief) and small (< 100 m relief) composite-ridges is somewhat arbitrary, but one significant difference does exist. Large composite-ridges generally incorporate a considerable volume of pre-Quaternary bedrock that is consolidated to some degree. Small composite-ridges may or may not include such bedrock; in fact, many are composed mostly of unconsolidated Quaternary strata. Being more susceptible to both glacial and nonglacial erosion, such ridges cannot maintain a high topographic relief. The main genetic difference between large and small composite-ridges, thus, is incorporation of consolidated bedrock.

The term *push-moraine* is commonly and loosely used in reference to ridges created by ice shoving. As applied here, push-moraines are a restricted subset of small composite-ridges that consists largely or wholly of glaciogenic strata. Composite-ridges that contain appreciable non-glacial material should not be called push-moraines. One common situation for push-moraines is thrusting of contemporaneous proglacial or ice-contact drift during continued glacier advance. Many push-moraines of this type were created in outwash deposits by glacier advances on Arctic islands of northeastern Canada during the Little Ice Ages (Kalin 1971).

The push-moraine of glacier C79 on southwestern Bylot Island (Map 1) is an example of a recent small composite-ridge (fig. 4–1). The distal portion consists of imbricately stacked scales of outwash sand and gravel. Each scale forms a low bench or terrace that slopes upglacier at a shallow angle (2–10°). Well preserved bedding within the scales dips at similar low angles, and the original stratigraphy of outwash beds is repeated in each thrust block. These distal ridges carry no trace of till and so must have been thrust in front of the ice margin. Wood twigs from the top of the innermost scale have yielded a corrected C–14 date of 120 ± 80 years BP, thus establishing the maximum age for ice pushing (Klassen 1982).

The higher proximal ridge stands 30–35 m above the modern sandur, is till covered, and marks the ice margin at the time of thrusting. The proximal ridge is

Fig. 4–1. Oblique aerial photographs of push-moraine in front of glacier C79, Bylot Island, Canada. Overview above (GSC 203639-I); close-up view below (GSC 203099-P). Proximal ridge to right is till covered (bouldery surface); distal ridges to left are tilted scales of outwash sand and gravel (smooth surface). From Klassen (1982, figs. 55.2 and 55.4). Copyright Geological Survey of Canada; published with permission of the Minister of Supply and Services Canada.

presumably also cored by thrust blocks of outwash strata. The scales of outwash were likely thrust while permafrozen. Low ridges forming a zigzag pattern in front of the push-moraine are probably patterned-ground fractures. The presence of these frost cracks may have controlled the size and initial development of the scales.

Small composite-ridges may be difficult to identify as glaciotectonic landforms and have often been mapped as end moraines. Many so-called end moraines are now recognized to consist partly or wholly of ice-shoved material (Moran *et al.* 1980). Where disturbed bedrock is present, the glaciotectonic origin of such moraines is obvious. However, the absence of deformed bedrock does not preclude ice pushing as the primary means of constructing certain end moraines.

Deformation of Quaternary strata by ice pushing is perhaps even more common than disruption of bedrock. Recognition of this situation can prove troublesome, however, as Moran (1971) pointed out. This is particularly true of push-moraines, where the disturbed glaciogenic strata may be similar in lithology to the enclosing drift. A complete transition is possible between conventional end moraines built by primary deposition and push-moraines created by secondary deformation. Only by careful examination of internal structures can push-moraines be identified properly.

Owing to their common occurrence and diverse nature, case examples are presented from quite different settings. Brandon Hills, Manitoba and Utrecht Ridge, the Netherlands are both Pleistocene examples in which pre-existing Quaternary strata were thrust in a manner much like large composite-ridges. Holocene push-moraines formed by glacier surges in unfrozen proglacial sandur sediment of Iceland and marine strata of Spitsbergen fjords are the third and fourth examples.

Brandon Hills, Manitoba, Canada

Brandon Hills are a small group of subparallel ridges, located at the northern end of Tiger Hills upland, about 11 km south of the city of Brandon (fig. 4–2). The Tiger Hills upland is part of the Manitoba Escarpment, located near the eastern edge of the Saskatchewan Plain. This escarpment rises abruptly as much as 300 m above the Manitoba Plain to the east, and is marked by a series of uplands – Duck Mountain, Riding Mountain, and Tiger Hills. These uplands are cored by upper Cretaceous shale of the Riding Mountain Formation. The bedrock of southwestern Manitoba is almost completely mantled by thick drift consisting mainly of till. Stratified drift is abundant on the uplands, however, where ice stagnated during deglaciation.

The glacial geomorphology and stratigraphy of southwestern Manitoba have been described in several investigations (Klassen 1975, 1979; Fenton, Moran *et al.* 1983; Fenton 1984). Brandon Hills have usually been mapped as part of the end moraine atop Tiger Hills upland (Prest *et al.* 1967).

Welsted and Young (1980), however, questioned designation of Brandon Hills as an end moraine, because they found that much of the hills consists of stratified

Fig. 4–2. Physiographic features of southwestern Manitoba. Location of Brandon Hills shown by solid dot at northern end of Tiger Hills upland. Adapted from Klassen (1979, fig. 1).

drift. They also considered, but rejected, the possibility of glacier thrusting apparently because deformed bedrock is not present. However, more recent study of Brandon Hills has shown that they are, in fact, ice-shoved ridges consisting entirely of drift (Aber 1988b).

Brandon Hills occupy a rectangular area roughly 10 km east-west and 4 km north-south (fig. 4–3). The hills exceed 590 m (1600 feet) elevation, up to 100 m above the Little Souris River valley immediately to the north. Brandon Hills include three distinct morphologic types: composite-ridges, esker ridges, and kame-and-kettle moraine (fig. 4–4). Closely spaced, subparallel, composite-ridges make up the northwestern, central and eastern portions of Brandon Hills. The ridges resemble a giant fishhook in overall plan view. The composite-ridges are covered by a thin veneer of till.

A high esker ridge meanders over the eastern end of Brandon Hills. This ridge extends with a couple of interruptions to the southwest and then northwest to form a large semicircular loop around the eastern half of Brandon Hills. The esker ridges are composed predominantly of sand and gravel and rest on a continuous substratum of till (Welsted and Young 1980). The southwestern flank of Brandon Hills consists of low kames and lake-filled kettles, such as Lake Clementi.

The internal structure of composite-ridges in Brandon Hills is fairly simple at most sites, consisting of 2–3 m of stratified sand and gravel overlain by 1–2 m of sandy till. The till is banded and is composed mainly of material eroded from subjacent stratified drift. Deformed stratified drift is visible in the largest and deepest exposure (site 6).

Site 6 is a large gravel pit cut into a small southeast-trending ridge. Pit walls

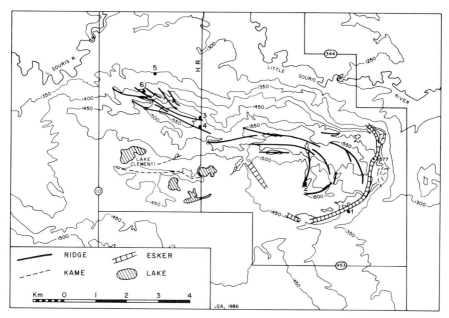

Fig. 4–3. Topographic map of Brandon Hills vicinity showing glacial features. Based on interpretation of aerial photographs. Provincial highways shown by numbered circles; H.R. = Hydraulic Road. Elevations in feet; contour interval = 50 feet (app. 15 m). Locations of exposures shown by solid dots. Taken from Aber (1988b, fig. 4).

reveal two dislocated stratified drift blocks, each 15 m thick, which are faulted together, tilted southwestward, truncated and capped by 1–2 m of discordant till (fig. 4–5). Both stratified blocks have a similar sequence; the lower portions of each consist of sandy pebble gravel with scattered cobbles, passing conformably upward into trough cross-bedded, medium to coarse pebbly sand.

Along the fault separating the upper and lower stratified drift blocks, trough-bedded sand of the lower block is either cleanly truncated or else sheared into a zone of foliated sand up to 0.5 m thick. This sand forms small apophyses which extend into the gravel above, and small thrusts and kink folds are present within this zone. Near the southwestern end of the section, a complex of high-angle to near-vertical faults cuts through sand of the upper unit, and the till/sand contact is offset here by small normal faults. The structural data are in complete agreement; faults all strike subparallel with the east-southeast-trending ridge.

The Lostwood (late Wisconsin) Glaciation completely covered Manitoba and reached far southward into the United States (Fenton 1984). Deglaciation was accomplished by stagnation of large areas of marginal ice alternating with minor readvances or stillstands of active ice. The Assiniboine Sublobe was responsible for the final readvance – Marchand phase – into Tiger Hills (fig. 4–6). This sublobe consisted of two ice tongues; one advanced directly over what is now the city of Brandon and the other moved into Tiger Hills. A re-entrant between the two ice tongues was developed near the eastern end of Brandon Hills.

Fig. 4-4. Aerial photograph of Brandon Hills vicinity, Manitoba. Composite-ridges are covered by forest; esker ridges are grass covered. Scale bar is 1 km long; north toward top. Aerial photograph A24519-179; copyright 1976. Her Majesty the Queen in Right of Canada, reproduced from the collection of the National Air Photo Library with permission of Energy, Mines and Resources Canada.

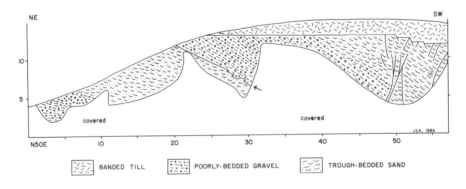

Fig. 4-5. Measured section from pit wall transverse to ridge axis at site 6, as it appeared in 1986. Scale in m; position of major fault zone shown by arrow.

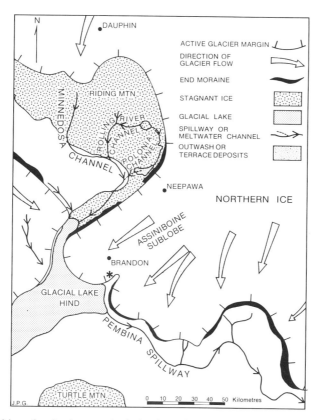

Fig. 4–6. Position of active ice margin during last advance into Tiger Hills upland. Note two ice tongues comprising Assiniboine Sublobe; position of Brandon Hills shown by asterisk. Adapted from McGinn and Giles (1987).

The creation of Brandon Hills is most easily explained by a single ice advance, which pushed up ridges consisting of displaced blocks of older stratified drift. This drift may have been deposited when ice stagnated over the Tiger Hills upland prior to the Marchand phase. Western and central ridges of Brandon Hills were shoved up along the southern flank of the Brandon ice tongue. Meanwhile, the Tiger Hills ice tongue pushed against the eastern end of Brandon Hills, causing ridges there to curve southward. The ice-shoved ridges were then overridden, and a veneer of till was deposited over the ridges. When ice advance ceased, a subglacial melt-water tunnel system developed, in which an esker system was deposited.

Ages given for the Marchand advance range from > 15,000 years BP (Klassen 1975) to about 11,200 years BP (Fenton, Moran *et al.* 1983). The discrepancy in ages is based on acceptance of selected radiocarbon dates. Teller *et al.* (1980) reviewed the complications of using radiocarbon dates from the Canadian Plains. The latest compilation of Laurentide ice-margin positions by Dyke and Prest (1987b) indicates Brandon Hills are between 13,000 and 12,000 years old. In any

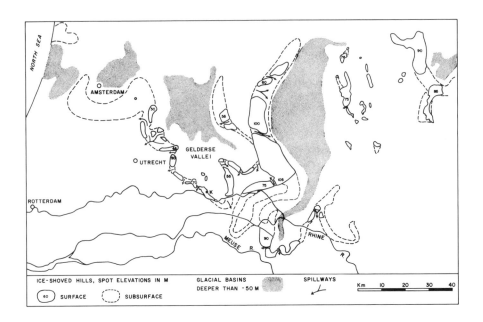

Fig. 4–7. Ice-pushed hills and glacial basins of the central Netherlands. K = location of Kwintelooijen sand pit within Utrecht Ridge. Based on van den Berg and Beets (1987).

case, Brandon Hills are an excellent example of the typical morphology and internal structures of small composite-ridges developed wholly within drift.

Utrecht Ridge, the Netherlands

Utrecht Ridge is located on the southwestern edge of Gelderse Vallei in the central Netherlands (fig. 4–7). It is one of a series of ridges composed of imbricately thrust scales that loops around the margin of a glacial basin located in the north-central part of Gelderse Vallei. Utrecht Ridge trends northwest-southeast, and it rises > 50 m above the adjacent lowlands, forming an impressive sight in an otherwise flat landscape.

Utrecht Ridge was created when an ice lobe excavated Gelderse Vallei basin. The ice-shoved hills are heavily eroded as a consequence of their Saalian age combined with unconsolidated character of Quaternary strata and are mere ruins of the original landforms. Kwintelooijen sand pit is located on the inner or northeastern side of Utrecht Ridge (fig. 4–7). The pit came under coordinated, interdisciplinary study during the 1970s and early 1980s on account of the many Paleolithic artifacts and fossils found there (Ruegg and Zandstra 1981).

The sedimentary strata now forming Utrecht Ridge were originally deposited as alluvium of the ancestral Rhine/Meuse Rivers. Kwintelooijen sand pit contains

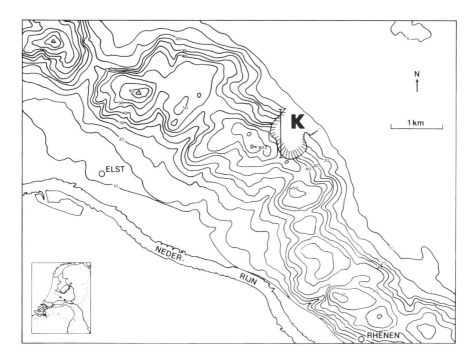

Fig. 4–8. Topographic map of Utrecht Ridge showing location of Kwintelooijen sand pit (K) and position of section in Fig. 4–9. Elevations in m; contour interval = 5 m. Modified from van der Wateren (1981, fig. 1).

three formations, in ascending order: Kedichem, Urk, and Drente (Ruegg 1981). The Kedichem Formation consists of very fine sand, clay, peat, and loam deposited in a flood-plain/back-swamp environment. Sediments of the Urk Formation are mainly fine to coarse sand, gravely sand, and coarse gravel deposited by the ancestral Rhine River (Zandstra 1981). The Drente Formation contains fine to coarse sand, gravely sand, and gravel. These formations total > 20 m in thickness.

The Drente Formation, by definition, includes all drift related to the Saalian glaciation. At Kwintelooijen, it accumulated as ice-margin and sandur deposits in front of glacier ice. Much of the Drente sediment was probably reworked from underlying Urk Formation deposits. In southern Gelderse Vallei, Saalian till is buried 15–30 m below sea level. A test boring made at the bottom of the sand pit penetrated thrust and contorted strata to a depth of about 24 m below sea level (Zandstra 1981).

Utrecht Ridge near Kwintelooijen is about 2.5 km wide with a plateau top between 45 m and 60 m above sea level (fig. 4–8). The ridge is asymmetric in cross profile. The northeastern side slopes 5–15°, whereas the southwestern flank slopes only 2–5°. This difference is partly explained by the presence of sandur deposits covering the southwestern portion of the ridge (van der Wateren 1981).

Utrecht Ridge and other ridges surrounding Gelderse Vallei are cut by several

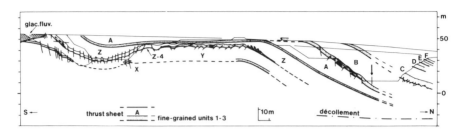

Fig. 4–9. North-south section through western portion of Kwintelooijen sand pit. Section trends obliquely to northwest-trending ridge. Individual thrust blocks lettered (X, Y, Z and A-F). See Fig. 4–8 for location of section. Taken from van der Wateren (1985, fig. 2).

dry valleys representing former spillways, and the ridge plateaus show no trace of till. The morphologic features demonstrate that Utrecht Ridge was not overridden by the Gelderse Vallei ice lobe (van der Wateren 1985); pushing of the ridge took place along the lateral margin of the ice lobe.

Utrecht Ridge consists of imbricated thrust blocks striking parallel to the ridge and dipping on average 35° to 40° NNE (fig. 4–9). Thickness of scales varies from about 25 m to only a few m, but each includes a basal portion of fine-grained sediment of the Kedichem Formation. Thrust blocks are imbricately stacked and gently folded. Thrusts at the base of each scale contain shear planes, small isoclinal folds, breccia and slickensides in a zone of intermingled sediments several dm thick. Many normal faults forming conjugate sets are also present.

The present elevation of Utrecht Ridge is less than its initial elevation, as a result of lowering by normal faulting and by later erosion. Van der Wateren (1985) estimated that scales were initially pushed up at least 100 m above the basal decollement. Structural restoration results in a good balance between volumes of Utrecht Ridge and the excavated basin of southern Gelderse Vallei.

Van der Wateren (1981) calculated that potential shear stress developed at the glacier sole would be far too small to upthrust blocks 100 m above the decollement. Instead of shear stress, the lateral pressure gradient caused by differential loading of the substratum during ice advance provided the driving force for glacier thrusting.

Five phases of Saalian ice pushing have been recognized in the Netherlands (fig. 4–10), three in the central portion and two in the northeastern region (Maarleveld 1953; ter Wee 1962). According to the traditional interpretation, the oldest phase (a) marks the maximum Saalian ice coverage, when Utrecht Ridge was built. Each younger phase represents a readvance of uncertain magnitude during Saalian deglaciation. During each of these phases, the ice margin was highly irregular, with ice lobes extending beyond the main inland ice sheet.

An alternate interpretation is based on increasing subsurface data from glacial basins (van den Berg and Beets 1987; de Gans *et al.* 1987). A series of deep glacial

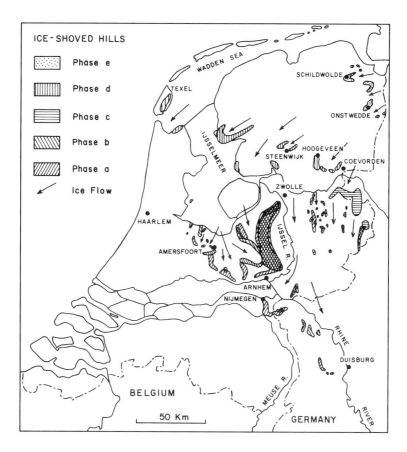

Fig. 4–10. Map of the Netherlands showing locations of ice-shoved hills and directions of ice movement. Taken from Aber (1985a, fig. 2).

basins behind ice-shoved hills stretches across the central Netherlands (fig. 4–7). The basins and associated hills increase in size from west to east. Owing to greater subsidence, ice-pushed ridges of the western basins are now mostly buried beneath younger sediments. These basins exceed 100 m depth and are floored with Saalian till in many places. The basins also contain tunnel valleys that appear to lead toward spillways breaching the marginal ice-shoved hills.

Prior to the Saalian glaciation, the Netherlands was an alluvial plain with very little topographic relief, having no major valleys or hills. The northern Netherlands was underlain by mostly fine-grained sediments over which the Saalian ice sheet advanced easily. Small ice-shoved hills (d, e) involving only the uppermost 10 to 20 m of preglacial strata were thrust, overridden, and streamlined during ice advance. As the ice sheet reached the central Netherlands, it overran coarser, gravely sediment (Urk Formation) of the ancestral Rhine/Meuse Rivers. This

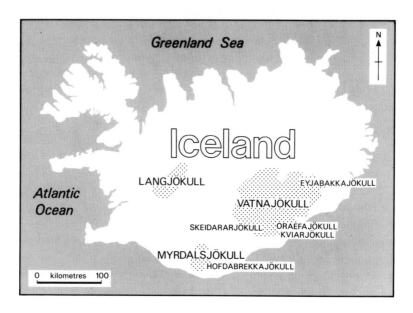

Fig. 4-11. Map of Iceland showing glaciers with recent, small composite-ridges.

caused an increase in basal friction, thickening of the ice sheet, and melt-water erosion of subglacial basins and tunnel valleys.

The marginal effect of this modified subglacial topography was to generate ice lobes in a region where the landscape was previously flat. The thicker ice sheet in combination with melt-water erosion of subglacial basins and development of ice lobes was an ideal situation for thrusting of composite-ridges (a, b, c) above a basal decollement in the Kedichem Formation. Two mechanisms contributed to thrusting: (1) lateral pressure gradient due to differential loading (van der Wateren 1985) and (2) direct glacier pushing against the sides of the basins (van den Berg and Beets 1987). Thus, the glacial basins were created by joint subglacial melt-water erosion and glaciotectonic thrusting.

Certain factors remain unresolved, for example the nature of permafrost. Most Dutch geologists have assumed that deformation took place in permafrozen sediments and that the thickness of scales is an indication of the depth of permafrost (de Jong 1967). Van der Wateren (1985), however, challenged the assumption of permafrost as a prerequisite for ice thrusting.

Jelgersma and Breeuwer (1975) explained the origin of glacial basins as deep erosion caused by glacial surging, and this is hardly compatible with ice advance over permafrost. Another possibility is that the ice sheet was cold-based during the advancing stage. Later, when the ice warmed and became more mobile, glacier lobes cut deep basins and pushed up ridges (ter Wee 1983).

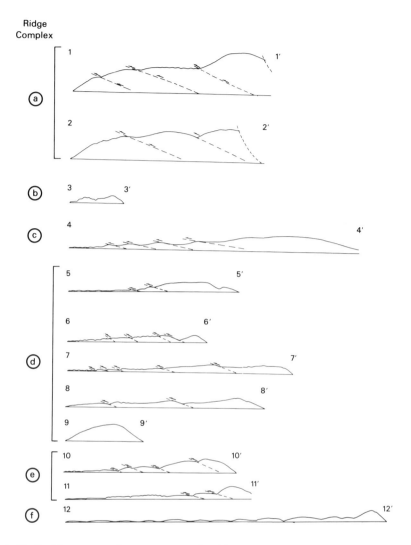

Fig. 4–12. Profiles surveyed across composite-ridges at the snout of Eyjabakkajökull, Iceland. Each profile is parallel to the direction of former ice movement (from right to left). Modified from Croot (1988b, fig. 4).

Vatnajökull, Iceland

Small composite-ridges are present at the snouts of several glaciers in Iceland: Höfdabrekkujökull, Kviarjökull, Skeidararjökull (fig. 4–11). Larger and more extensive composite-ridges of very recent age occur at the snout of Eyjabakkajökull, an outlet glacier of Vatnajökull (Croot 1978, 1987; Sharp 1982; Chapter 2). These ridges mark the ice limit achieved by the Holocene maximum advance of A.D. 1890. The Eyjabakkajökull ridges were formed by rapid ice advance into

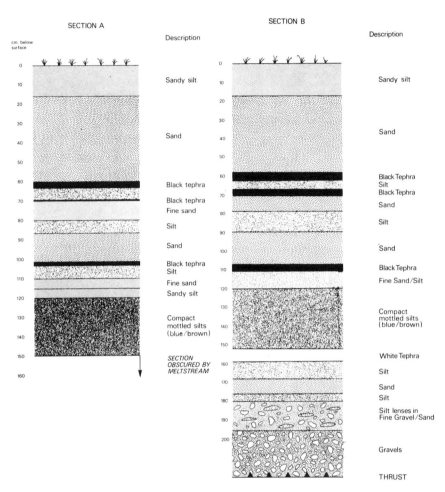

Fig. 4–13. Sections in proglacial sediments disturbed during a recent (1890 A.D.) advance demonstrate stratigraphic continuity from undisturbed (A) to disturbed (B) portions. Modified from Croot (1988b, fig. 6).

proglacial sandur deposits. The ridges occur as seven discrete sets, each crescentic in plan, but of varying size and shape (figs. 2–12 and 2–13).

Each set represents a unit of deformation that was produced by an individual segment of the ice lobe and is probably separated from the rest by longitudinal shear zones (fig. 2–7). Except for the westernmost set (a in fig. 2–13), which is more eroded, the ridge sets have retained their original morphology. The dimensions of each ridge set vary, but they have a common form. The ridge crest nearest the former ice margin is the highest, and crest heights fall progressively outward (fig. 4–12). The larger proximal ridges are asymmetrical with steep slopes facing downvalley and more gentle slopes facing in toward the former glacier margin. The small distal ridges are symmetrical.

Fig. 4–14. Ground-photo mosaic of composite-ridges at the snout of Eyjabakkajökull. Ice movement was from left to right, but the glacier only reached as far as the left-hand edge of the photograph. The continuity of vegetation and turf from ridges to undisturbed sandur is evident. Deformed turf comprises the bulk of the largest ridges (extreme left of photo).

River-bank sections demonstrate the stratigraphic continuity from undisturbed sandur into the ridges (fig. 4–13). This is supported by the continuity of the turf and near-surface stratigraphy elsewhere in the ridges and sandur. Tephra layers provide excellent marker horizons. The remainder of the strata are sand, silt, and gravel beds. Some of these sediments are loose-textured, but others are compact and impermeable.

Prior to the 1890 advance, the area had developed an extensive vegetation cover, with a substantial depth of turf, organic soil, and peat in some places. This material was incorporated into the push-moraine sets and comprises the bulk of the melange that forms the core of the largest ridge at the former ice margin (fig. 4–14). Well-defined thrust sheets are expressed at the surface as asymmetrical ridges. Farther beyond the former ice margin, the smaller asymmetrical ridges are seen to be simple anticlines or overturned anticlines. In this way a straightforward relationship between topography and underlying structures is evident.

The overall section may be divided in two portions characterized by contrasting styles of deformation: (1) subglacial portion beneath the former (1890) ice margin and (2) proglacial portion that lay downvalley from the ice margin (fig. 4–15). The main part of the subglacial portion is marked by a number of low-angle normal faults dipping in the direction of ice movement (transport direction).

The amount of slip, as shown by a displaced tephra bed, ranges from 0.5 m to just less than 6 m with no apparent systematic variation. The downvalley dip of the faults increases in the direction of transport from 4° to 17°. The arrangement of the fault set implies convergence of faults at a depth of 4–5 m, some 10 m inside the former ice margin. Gravel below the lowest visible fault appears to be compacted, but undisturbed stratigraphically, and may represent the basal decollement. In addition to these low-angle normal faults, an upper roof thrust has carried horizontally bedded gravel and a mixture of till and gravel downvalley.

Between these subglacial structures and those which developed beyond the former ice margin, a transitional zone is found, in which substantial deformation

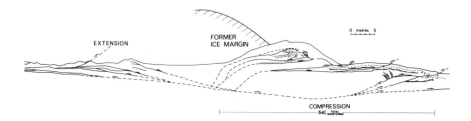

Fig. 4–15. Section parallel to ice movement through composite-ridge set e (fig. 2–13), showing subdivision of the ridge in two broad elements: subglacial and proglacial.

apparently formed partly beneath and partly in front of the glacier. This includes the largest ridge (fig. 4–16). The ridge comprises at least three sheets of extremely disturbed silt and soil bounded by well-defined thrusts. The original stratigraphy, as indicated by tephra layers, appears to have been identical to that found immediately downvalley.

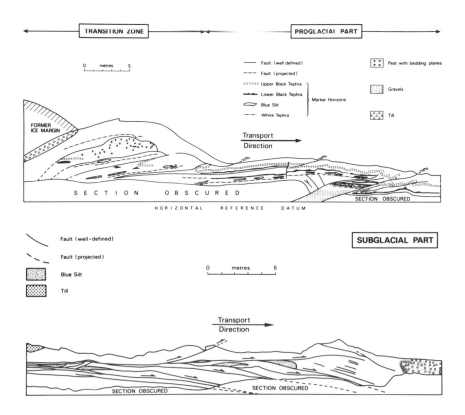

Fig. 4–16. Detailed enlargements of Fig. 4–15 showing change in nature of deformation from subglacial to proglacial portions. Adapted from Croot (1988b, figs. 8–9).

Most of the proglacial portion displays quite different traits from the subglacial portion. The whole section is divisible into several imbricated thrust sheets dipping upvalley, opposite to the direction of transport. The majority of thrusts are concave upward in their upper parts and convex upward lower down. The thrusts are all low angle (3–20°), the majority reaching the surface.

The amount of displacement, as shown by tephra beds, along three of the major proglacial thrusts varies from < 3 m to > 8 m (fig. 4–16). Some thrusts die out upward into fold structures, particularly farthest from the former ice margin. Folding is mainly associated with the upper horizons, especially in the distal area in the toe of thrust sheets. Some normal faults are developed in the strata, displacing the major thrusts by a few cm.

Prior to deformation, the stratigraphy throughout the proglacial portion appears to have been similar to that in the undisturbed sandur, and restoration extends this stratigraphy as a planar surface some 50–70 m farther upvalley (fig. 4–17). As Eyjabakkajökull surged to within 50–70 m of the 1890 limit – the vicinity of lake-filled depressions, it detached sheets of turf and silt from the gravels beneath.

Such detachment was facilitated by high pore-water pressure generated within the gravels. A natural plane of decollement developed either within the gravel or at the base of the overlying beds. The measured rates of conductivity (K) for undisturbed silts (0.0047 cm/s) and gravels (0.172 cm/s) demonstrate the contrasting abilities of these materials to transmit water (Croot 1987).

Recent measurements of melt water production during surging show that water is normally generated in large volumes at the glacier base (Kamb et al. 1985). Provided this melt water permeates the substratum and becomes linked with ground water in the proglacial area, then any melt water generated at a rate in excess of the conductivity of the gravels would induce significant pore-water pressure within the confined gravels. The overburden stress imposed by rapidly advancing ice would have added significantly to the rate of loading (van der Wateren 1985).

As the sheets of silt and turf became imbricated and were pushed along the surface in front of the advancing ice margin, excess pore-water pressure continued to be generated within the gravel beds under the transition zone and proglacial portions of the composite-ridge. A synchronous movement then appears to have taken place involving both subglacial faulting and proglacial thrust movements. The two systems appear to be physically linked by a continuous decollement. Although only small portions of this decollement appear in the section, the projection of major faults in both the subglacial and proglacial portions indicates its presence.

Relationships of thrusts in the proglacial portion suggest that they developed in a sequential manner (fig. 4–17). The upper beds appear to have behaved as competent strata, whereas the thrusts propagated within gravel and cut upsection through the silt units above. The numerous channels superimposed across the trend of the ridges originate at the outcrops of thrust planes. This indicates that the thrusts were water-lubricated during movement and were the sources of springs for the release of ground water trapped under high pressure.

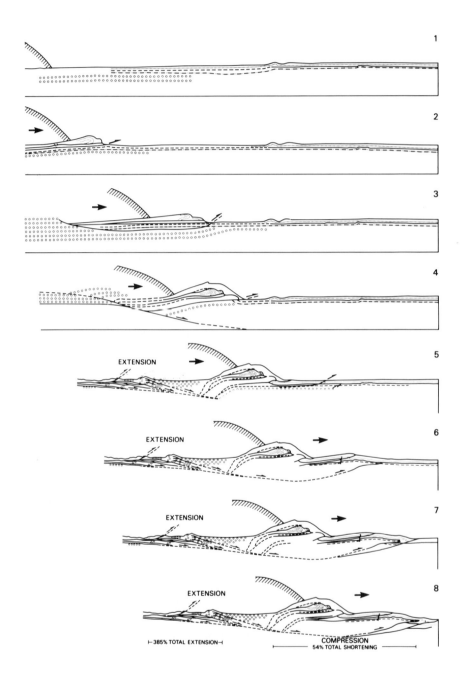

Fig. 4–17. Sequential development of ridge set e at the snout of Eyjabakkajökull, Iceland, as interpreted from post-deformation data. Taken from Croot (1988b, fig. 11).

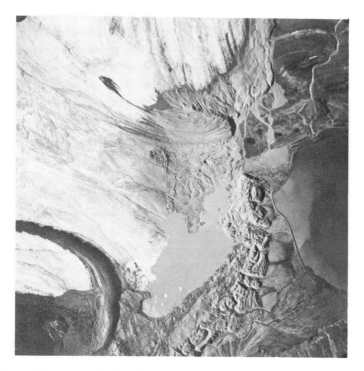

Fig. 4–18. Aerial photograph of small composite-ridges at the snout of Gronfjordbreane, a surging glacier in Svalbard. Reproduced with permission of Norsk Polarinstitutt.

During ice advance, as stress increased, strain was initially accomplished by folding, followed by thrusts propagating close to the advancing glacier snout. As each thrust movement waned, the next thrust took up the movement, initially by ductile folding, but subsequently by failure along the gravel and cutting upsection through the silts and turf. The final movements involved progressively lower levels of stress, and led to concealed thrusts and eventually to simple fold structures at the toe of the system.

Spitsbergen, Svalbard, Norway

Reports of Jura-like ridges in front of glaciers in Spitsbergen date back as far as the 1890s (Garwood and Gregory 1898) with more detailed observations early in this century (Lamplugh 1911; Gripp and Todtman 1925; Gripp 1929). The increased availability of aerial photographs and satellite images has led to recognition of the common occurrence of composite-ridges at the snouts of glaciers in Svalbard (fig. 4–18). Many such ridge sets fill the entire width of the proglacial area in front of some glaciers, whereas other sets occupy smaller portions. Ridge development is favored by rapid glacier advances into low-lying sandur or fjord-head regions.

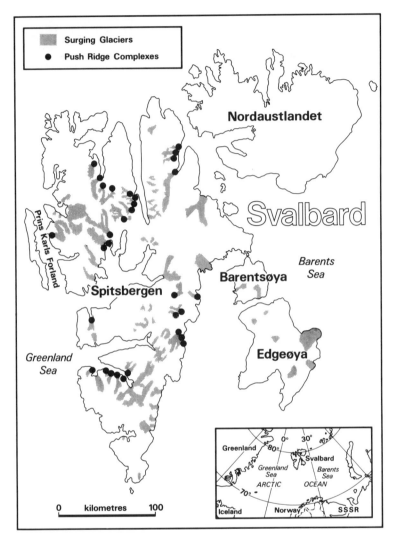

Fig. 4–19. Correlation of surging glaciers and composite-ridges (push-moraine complexes) in Spitsbergen, Svalbard. Taken from Croot (1988c, fig. 4).

Croot (1987) suggested a strong association between composite-ridges and surging glaciers (fig. 4–19). Establishing such a correlation is difficult, however. Even those glaciers that surge regularly may not do so at intervals more frequently than 50–100 years. The earliest reliable accounts of rapid advances of the more accessible Spitsbergen glaciers date back only as far a 1890. Consequently indirect evidence must be used in many cases. In many instances, morphologic evidence derived from air photos or satellite images can be used to establish a history of surging (Meier and Post 1969). In a small minority of cases, where composite-ridges occur at a glacier snout, glacier behavior remains unknown.

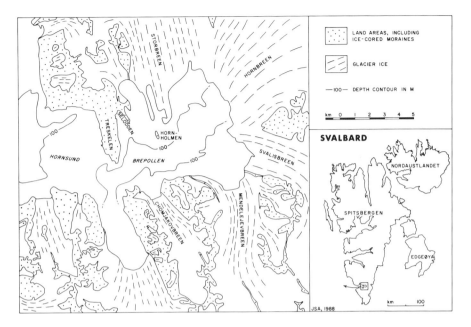

Fig. 4–20. Map of inner Hornsund vicinity, southern Spitsbergen, showing location of Treskelen push-moraine at edge of Brepollen basin. Based on Karczewski (1984); ice-margin positions of 1983.

The inland, protected fjords of Spitsbergen are rapidly filled by high rates of sediment discharge from melt water streams. The upper 5–8 km of these fjords is typically a shallow tidal shoal system. These tidal flats grade gently into proglacial environments characterized by sand, silt and clay deposits, which are ideal sediments for glaciotectonism. Where glaciers advanced into steeper, sloping sandur or those characterized by coarse-grained sediments, composite-ridges are usually absent.

Treskelen (the Threshold) is a good example of a tidewater push-moraine from the inner portion of Hornsund, southern Spitsbergen (fig. 4–20). Treskelen is located on a shallow (< 60 m deep) sill beneath Hornsund that partly isolates the deeper (140 m) basin of Brepollen at the head of the fjord system. Treskelen, along with Selodden and possibly Hornholmen, were shoved up when the combined glaciers of Storbreen, Hornbreen, Svalisbreen, Mendelejevbreen and Chomjakovbreen advanced across Brepollen during the Little Ice Ages.

The crest of Treskelen consists of two or three parallel ridges that are ice-cored moraines (fig. 4–21). The proximal (eastern) side is made up of debris covering dead ice, and the distal side has outwash sediment resting on dead ice (Karczewski 1984). Selodden is covered by ground moraine.

Push-moraines of Treskelen and Selodden contain clayey glaciomarine strata that were presumably shoved out of the Brepollen basin. Various fossil shells have

Fig. 4–21. Aerial photograph of Treskelen, inner Hornsund, and adjacent tidewater glaciers. Scale bar is 1 km long; north toward top. Copyright Norsk Polarinstitutt; S70 4654, 23 Aug. 1970.

given radiocarbon dates in three age ranges: (1) 9770–8400 years BP, (2) 4200–4000 years BP, and (3) about 2000 years BP (Birkenmajer 1987). In addition, a piece of driftwood from the push-moraine was dated at 810 years BP. These dates indicate that inner Hornsund was ice free with accumulation of marine sediment taking place throughout most of the Holocene. The ice advance which built Treskelen and Selodden occurred during the last few hundred years.

It is not known if Treskelen and Selodden were built by surging glaciers or by normal ice advance; however, surging is a common form of glacier advance in Spitsbergen (Elverhøi *et al.* 1983). With so many glaciers entering the head of Hornsund, it seems probable that one or more of them did surge when the push-moraines were built. A portion of the push-moraine material was shoved up from below sea level, which means that the uplifted sediment was surely unfrozen at the time of deformation.

CHAPTER 5

CUPOLA-HILLS

Introduction

Many conspicuous hills, both small and large, have the general characteristics of ice-thrust masses, but lack the hill-hole relationship or the typical ridged morphology. Bluemle and Clayton (1984) placed such hills from North Dakota into the category of *irregular hills*, that Clayton et al. (1980:46) defined as 'an irregular jumble of hills with no obvious transverse ridges and no obvious source depression.'

The glaciotectonic origin of such hills can only be proven with evidence of subsurface deformation of bedrock or drift, although such hills may be suspected from other evidence. Ice-shoved hills of irregular shape are probably much more common than heretofore recognized.

Perhaps the most typical form of these irregular hills is the type that Smed (1962) first called Cupola-hill (*kuppelbakke* in Danish). Cupola-hills have an internal structure similar to composite-ridges; however, unlike composite-ridges their morphology was substantially modified by the action of overriding ice. Cupola-hills possess three basic attributes (Smed 1962):

1. Interior structure – deformed glacial and interglacial deposits, plus detached floes of older strata or bedrock.
2. External form – long, even hill slopes with overall dome-like morphology, unlike marginal moraines; varying from near circular to elongated ovals in form; 1–15 km maximum length, most 5–10 km long, from about 20 m to > 100 m high.
3. Discordant till – overridden by ice which truncated deformed structures and laid down a basal till layer.

The internal structure of cupola-hills is due to ice shoving of pre-existing strata during glacier advance; whereas, the external form reflects the smoothing effects of later subglacial erosion and deposition. Where subglacial modification is slight, a subdued ridge morphology may be preserved. With increasing subglacial erosion and deposition, all traces of the ridges disappear and are replaced by a smoothed cupola form. The long axis of a cupola-hill may be either parallel or perpendicular to ice movement. Still further subglacial molding may create streamlined, drumlin-shaped hills.

Eventually, the morphologic expression of earlier, ice-shoved hills may be completely altered or removed by prolonged glaciation. Thus, a complete transition from composite-ridges, to cupola-hills, to streamlined hills and plains is possible.

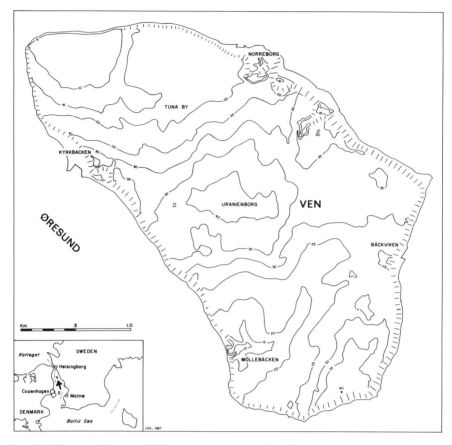

Fig. 5–1. Topographic map of Ven, southwestern Sweden. Contour interval = 5 m; location of Ven shown by arrow on inset map.

Cupola-hills are common in regions having soft substratum that were subjected to multiple glacial advances from different directions.

The Swedish island of Ven, located in the Øresund strait between mainland Sweden and Denmark, is a good introductory example of a typical cupola-hill (fig. 5–1). The island is 4.5 km long by 2.4 km wide and reaches a maximum elevation of 45 m at Uranienborg. Uranienborg is situated on a subdued ridge that trends across the center of the island. From this central ridge, the land slopes smoothly and gently toward the north and south, and a second smaller ridge is present at the southern tip of the island. Coastal erosion has cut steep cliffs along the island's eastern, southern and western sides.

The island consists of glaciogenic formations that were deposited, deformed, and overridden during a series of late Weichselian ice advances from north to northeast and from east to southeast (Adrielsson 1984). The major structures are gentle synclines, some 400–500 m across and 20–30 m in amplitude, that are disrupted by numerous thrust faults and smaller overturned folds. Discordant tills cap the entire island, except where removed by postglacial erosion.

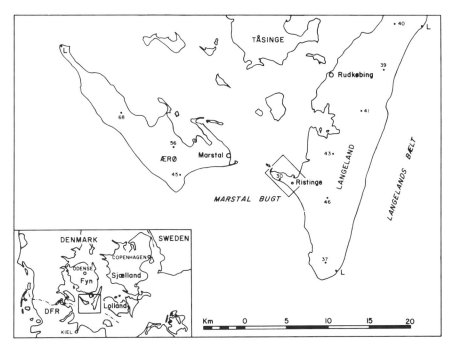

Fig. 5–2. Location map for Ristinge, Langeland, Ærø and surroundings, southern Denmark. Location of aerial photograph (fig. 5–4) shown by rectangle at Ristinge. Spot elevations on high points in m; L = lighthouse.

The base of structural disturbances apparently coincides with the boundary between the glaciogenic strata and underlying fluvial deposits, approximately 30 m below sea level. The sandy fluvial deposits are a major aquifer, in which elevated pore-water pressure may have developed during glaciation and thus facilitated deformation of overlying strata (Adrielsson 1984).

The following case examples range from a small drumlin-shaped hill at Ristinge Klint, southern Denmark to the large cupola-hill at Gay Head, Martha's Vineyard, Massachusetts. The former is composed entirely of Quaternary strata, whereas the latter consists mostly of Cretaceous and Tertiary bedrock material. The cupola-hill at Hvideklint, Møn, Denmark is intermediate in size and contains roughly equal amounts of deformed chalk and drift.

Ristinge Klint, Langeland, Denmark

A small cupola-hill forms the main part of Ristinge peninsula on the island of Langeland, southern Denmark (fig. 5–2). This cupola-hill is not topographically prominent; its high point at 30 m is well below hill-top elevations on the rest of Langeland and on nearby Ærø. The significance of Ristinge is found in the

Fig. 5-3. Photograph of Ristinge Klint as seen from the beach to the southeast. Cliff displays stacked scales of Quaternary strata. Height of cliff approx. 25 m; photo by J. Aber, 1987.

Quaternary sequence exposed on the southern side of the peninsula in Ristinge Klint (fig. 5-3).

In the 1-km-long cliff, more than 30 scales consisting entirely of Quaternary deposits are imbricately thrust with consistent southeasterly dips. The scales include late Saalian, Eemian, and Weichselian strata that are discordantly covered by late Weichselian and postglacial deposits. The fossiliferous marine Eemian beds are perhaps the best known of the various Quaternary strata.

Based on early geological investigations by Madsen *et al.* (1908) and Madsen (1916), Ristinge Klint has become a classic locality for late Quaternary stratigraphy. Ristinge was mentioned along with cupola-hills on Ærø and Fyn by Smed (1962). The cupola-hills on Ærø and at Ristinge have drumlin forms (fig. 5-4). High points of each are toward the southeastern ends with the hills tapering northwestward. These hills are the largest drumlins in Denmark, but their internal structure is hardly typical of drumlins. Smed (1962) concluded that ice pressure from the southeast had dislocated and streamlined these cupola-hills.

Fig. 5–4. Aerial photograph of Ristinge peninsula with Ristinge Klint on the southern (left) side. Note elongated, streamlined form of hill. Scale bar is 500 m long; north shown by arrow. Copyright Geodætisk Institut, Denmark, 1987 (A. 29/88).

More recent and comprehensive work at Ristinge Klint has been carried out by Ehlers (1978) and Sjørring *et al.* (1982), who described four Weichselian tills and one reworked late Saalian till along with intervening stratified deposits (figs. 5–5,

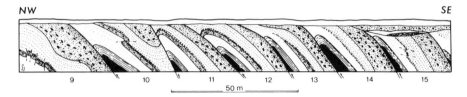

Fig. 5–5. Central portion of Ristinge Klint showing stacked scales of Quaternary strata. Numbering of scales after Madsen *et al.* (1908). For stratigraphy see Fig. 5–6. Taken from Sjørring *et al.* (1982, fig. 1).

5–6). Based on compositional traits and stratigraphic positions, the Weichselian tills are assigned as follows (Berthelsen 1973; Sjørring 1983; Houmark-Nielsen 1987):

1. Bælthav and East Jylland Tills – deposited by Young Baltic advances from southeast.
2. Mid Danish Till – related to Main Weichselian advances from the northeast.
3. Ristinge Klint Till – laid down by Old Baltic advance from southeast.

Ristinge Klint presents a repetition of scales, each up to 20 m thick and each containing essentially the same Quaternary sequence, imbricately stacked in a remarkably consistent structure (fig. 5–5). Shiny clay or Eemian clay forms the base of each scale, indicating that these clay formations were the zone of failure in which thrust faulting was initiated. The internal stratification of each scale is little disturbed, although some foliation and drag folding is developed adjacent to thrust faults. Structural measurements along the cliff show the average direction of upthrusting toward 322°, corresponding to ice movement from the southeast (Sjørring *et al.* 1982).

The discordant tills lie with angular unconformity on the tilted scales and are not involved in any way with the thrusting. Therefore, thrusting of scales and deposition of discordant tills were separate events. Thrusting presumably took place near the ice margin during glacier advance, and truncation of scales and till deposition happened during later ice overriding. Although discordant tills were locally deposited, the main result of overriding ice was erosion and streamlining of the cupola-hill into a large drumlin.

It is clear that the Young Baltic ice lobe moving from the southeast was responsible for thrusting at Ristinge Klint. The problem is how to interpret the two discordant tills of the Young Baltic advances. Sjørring *et al.* (1982) assigned the lower till and hence thrusting of Ristinge Klint to the initial East-Jylland phase of the Young Baltic glaciation. The upper till was assigned to the later Bælthav advance. The East-Jylland advance overspread all the islands, whereas the Bælthav advance was confined mainly to straits between larger islands, but did cover all of Langeland and Ærø. Another interpretation was given by Houmark-Nielsen (1981, fig. 12), who showed thrusting of Ristinge Klint as a result of Bælthav ice advance.

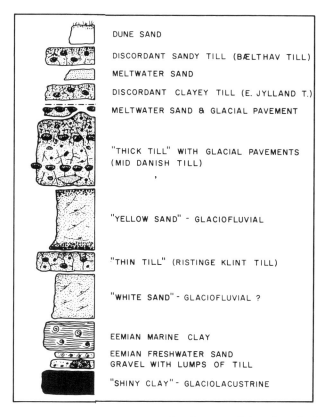

Fig. 5–6. Composite stratigraphic sequence for Ristinge Klint. One Saalian and four Weichselian tills are present. Adapted from Sjørring *et al.* (1982, fig. 2); formal till names given in parenthesis (from Houmark-Nielsen 1987).

Any permafrost existing in southern Denmark prior to deposition of the Mid Danish Till was probably removed by subglacial melting and erosion. Between the Main Weichselian and Young Baltic phases, a warm interval, the Asnæs interstade, was a time of permafrost melting, and likewise the Røsnæs interstade was a warm interval between the East-Jylland and Bælthav ice advances (Berthelsen 1975). Some thin permafrost might have locally developed in front of advancing ice lobes near the end of these interstades, but whether this was sufficient to control thrusting of floes at Ristinge is unknown. Thrusting within the clay decollement zone was likely facilitated by increased hydrostatic pressure as a result of glacier-ice loading.

Gay Head, Martha's Vineyard, Massachusetts, United States

Martha's Vineyard is one of several offshore New England islands built mostly of two prominent late Wisconsin end-moraine systems (fig. 5–7). These moraines are

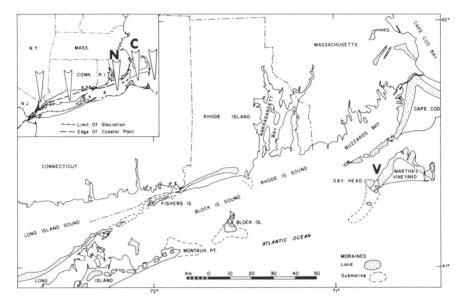

Fig. 5–7. Map of southern New England showing late Wisconsin end moraines: V = Vineyard Moraine. Inset map shows edge of Coastal Plain, approximate limit of late Wisconsin glaciation, and positions of ice lobes: N = Narragansett-Buzzards Bay and C = Cape Cod Bay lobes. Moraines from Schafer and Hartshorn (1965) and Sirkin (1980).

largely glaciotectonic features formed by advancing ice fronts (Oldale and O'Hara 1984), which piled up glacial, interglacial, and preglacial strata along with thrust and folded floes of Tertiary and Cretaceous bedrock. Gay Head Cliff, at the western tip of Martha's Vineyard, has long been famous for its exposures of deformed bedrock and drift. The cliff reaches 45 m in height and extends 1 km north-south. It cuts through the Gay Head cupola-hill of about 10 km². A close structural similarity between Gay Head Cliff and Møns Klint has been pointed out often.

Upthrust Cretaceous and Tertiary floes of Gay Head Cliff represent the glacially modified edge of the Atlantic Coastal Plain. Topographic troughs in the Precambrian and Paleozoic basement surface channeled ice flow resulting in more-or-less parallel ice lobes at the glacial margin. Martha's Vineyard occupies the interlobate position between the Narragansett-Buzzards Bay and Cape Cod Bay lobes (fig. 5–7).

Cape Cod and all the offshore islands are underlain by Cretaceous and Tertiary sedimentary strata of the Coastal Plain. This bedrock normally lies some depth below sea level and dips gently seaward. All surface exposures of Cretaceous or Tertiary material on Martha's Vineyard and other islands consist of deformed floes or thrust scales.

The late Wisconsin ice margin of southern New England was grounded on land, and the lobes apparently oscillated at times, but not simultaneously (Sirkin 1976, Oldale and O'Hara 1984). Thrusting of the Vineyard moraine took place near the

Fig. 5–8. Early cross section of Gay Head Cliff: A = upper Cretaceous, B = Miocene and Pliocene, C = older Pleistocene, D = thrust planes and faults. First published by Woodworth (1897); taken from Woodworth and Wigglesworth (1934, fig. 10).

end of glaciation, as considerable drift is involved in the deformations, but only a thin discordant till was deposited over the top of the truncated structures. The Vineyard moraine cannot be regarded as a terminal moraine, because it was locally overridden by ice. However, the distance of overriding was probably only a few km.

Different interpretations of the structure and stratigraphy of deformed bedrock and drift at Gay Head Cliff have generated many geological controversies over the years. In the 18th and 19th centuries, tectonic, landslide or volcanic activities were favored explanations for Gay Head Cliff, and some geologists promoted the concept of a post-Miocene 'Vineyard orogeny' (Shaler 1888, 1898). By the late 1800s, however, most geologists had accepted a glaciotectonic origin for the structure (Merrill 1886b; Hollick 1894; Woodworth 1897; Upham 1899), in which limbs of overturned folds and thrust faults dip northeastward (fig. 5–8).

TIME UNITS	LONG ISLAND	RHODE ISLAND and CONNECTICUT	CAPE COD, MARTHA'S VINEYARD, and NANTUCKET
late Wisconsinan	Roslyn till and outwash	New Shoreham drift	All drift on Cape Cod All drift atop the "Montauk till" All drift atop upper Sankaty Sand
middle Wisconsinan	Thrusted oyster reefs, peat deposits, and clay north of the Harbor Hill moraine, Western Long Island	Silts atop Montauk drift	
early Wisconsinan	Manhasset Formation including its Montauk Till Member	Montauk drift	"Montauk till"
Sangamonian	Tabaccolot fauna on Gardiners Island		Fauna of upper part of Sankaty Sand
	Westhampton Beach fauna Bridgehampton fauna		Fauna of lower part of Sankaty Sand
Illinoian or older			Nebraskan drift Aftonian deposits and Kansan drift. Drift below lower part of Sankaty Sand

Fig. 5–9. Stratigraphic correlation chart for glacial and interglacial deposits of southern New England and New York islands. Taken from Oldale (1980, Tab. 4); reprinted by permission of Kendall/Hunt Publishing Company.

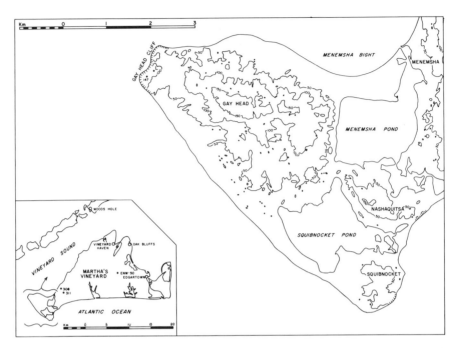

Fig. 5–10. Topographic map of Gay Head vicinity. Elevations in feet; contour interval = 50 feet (approx. 15 m). L = lighthouse atop Gay Head Cliff.

Upper Cretaceous, Tertiary, and Pleistocene formations make up the disturbed preglacial strata of Martha's Vineyard. These are all dominantly clastic sediments weakly indurated to non-indurated in character. Tertiary strata are normally encountered about 20–25 m below sea level and consist of glauconitic sand. The Tertiary/Cretaceous boundary lies about 70 m below sea level (Hall *et al.* 1980). The bulk of bedrock masses exposed at Gay Head Cliff consists of upper Cretaceous strata with white kaolinitic beds being especially prominent. In addition, two other preglacial formations outcrop in the cliff – Miocene greensand and lower Pleistocene Aquinnah conglomerate.

The traditional sequence for glacial and interglacial stratigraphy on Martha's Vineyard was developed in line with four major Pleistocene glaciations of the mid-continent region. Among the several glacial formations, the Montauk Till of the Manhasset Formation is the most prominent and is involved in deformations on Martha's Vineyard. Believing the Manhasset Formation to be early Illinoian in age, Kaye (1964a) assigned the main ice thrusting to a late Illinoian glaciation.

The Manhasset Formation is now regarded as no older than early Wisconsin (Schafer and Hartshorn 1965), because the underlying marine interglacial strata (Jacobs, Gardiners, Sankaty) are almost surely Sangamon or early Wisconsin in age (fig. 5–9). Thrusting of the Vineyard moraine (including Gay Head) is, therefore, late Wisconsin, because the moraine contains deformed Montauk Till (Oldale

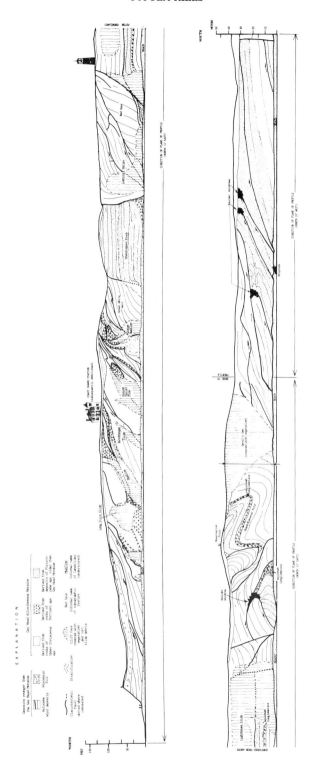

Fig. 5-11. Gay Head Cliff section as it appeared in 1959. Vertical scale = horizontal scale; base of section is sea level. Taken from Kaye (1980).

1980). Radiocarbon dates from the moraine in the 12,000 to 15,000 years BP range support this interpretation.

The cupola-hill of Gay Head is oval in shape, roughly 5 km long and 2 km wide, extending northwest-southeast (fig. 5–10). The cupola-hill is not ridged, but has instead uniformly sloping sides, on which there are many small, irregular knobs and depressions. Maximum elevation reaches nearly 60 m at the center of the cupola-hill in the town of Gay Head, some 75 m above the floor of Vineyard Sound to the north. Two smaller cupola-hills are located to the southeast at Nashaquitsa and Squibnocket.

The most detailed sketch of Gay Head Cliff, made by Kaye in 1959 (fig. 5–11), shows a series of imbricated scales, 20–30 m thick each, dipping toward the northeast. This general scheme is interrupted in places by complex folds or downdropped blocks. Beginning at the northern end, half a dozen large scales culminate southward with the Great White Wall (just east of Coast Guard station, fig. 5–11). These scales are comprised mainly of Cretaceous strata in their lower portions, and strongly folded Tertiary and Pleistocene material is present toward the tops of some.

The Great White Wall is faulted against a downdropped block of nearly horizontal drift resting on Tertiary bedrock at the base of the cliff. Immediately south of the down-dropped block, there is a large mass of steeply tilted and thrust Cretaceous strata, in which lignite is conspicuous. Continuing southward, nearly horizontal overthrusts are visible at Lighthouse Slide, followed by a complexly deformed zone in which the Quadrilateral Fold is developed.

The southernmost portion of the cliff (south of Devil's Den) consists of low-angle overthrusts involving Cretaceous strata and drift. Overall, the consistency in structural development suggests that a single northeasterly ice advance thrust up scales and subsequently overrode the Gay Head cupola-hill.

Northeast of Menemsha, elevations on the main portion of Martha's Vineyard reach a maximum of 95 m (311 feet, fig. 5–10). This area is marked by northeast-trending ridges, composed of Montauk Till, and Cretaceous strata striking northeast are uplifted at least 160 m above their normal undisturbed position. The Menemsha scales were apparently thrust up from the northwest resulting in large composite-ridges.

The Menemsha ridges along with Gay Head, Nashaquitsa, and Squibnocket cupola-hills outline an ice tongue, which advanced from the Menemsha Bight southward through the depression of Menemsha and Squibnocket Ponds. The same advance eventually overran the cupola-hills, but may not have covered the highest Menemsha ridges, which rose above the ice as nunataks (Kaye 1964b).

The Menemsha ice tongue was an offshoot of the Narragansett-Buzzards Bay ice lobe. Comparison of the Vineyard moraine with other coastal moraines of Massachusetts, Rhode Island, and New York indicates that most of these so-called end moraines may have had a similar glaciotectonic origin (Oldale and O'Hara 1984). The moraine systems actually consist of large and small composite-ridges and cupola-hills created by ice pushing.

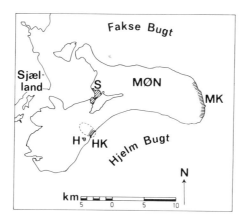

Fig. 5–12. Map of Møn, southeastern Denmark. HK = Hvideklint, H = village of Hjelm, S = city of Stege, MK = Møns Klint. Dashed line shows position of cupola-hill through which Hvideklint is cut. Modified from Aber (1979, fig. 1).

The New England coastal end moraines were formed by ice advance against the Coastal Plain cuesta and against ice-contact stratified drift. Floes of bedrock and drift were sheared off and transported forward and upward beyond the ice front. Structural uplift of at least 160+ m occurred on Martha's Vineyard. This thrusting was facilitated by impermeable silt and clay beds, within both bedrock and drift strata, where high pore-water pressure could develop during ice advance.

Gustavson and Boothroyd (1987) compared the late Wisconsin glaciation of coastal New England with the modern Malaspina Glacier of Alaska. They concluded that the Laurentide Ice Sheet of southern New England was a temperate ice mass, in which a large volume of melt water discharged from fountains or subglacial streams at the ice margin. Under such conditions, excess ground-water pressure could have developed in aquifers beneath and beyond the ice margin and led to thrusting of largely unfrozen sediment and bedrock.

Hvideklint, Møn, Denmark

The hills north of the village of Hjelm, including Hvideklint, on the southern coast of Møn island (fig. 5–12) are an ideal example of a cupola-hill built of displaced bedrock and drift. At Hvideklint, several upthrust floes of upper Cretaceous white chalk, covered and separated by deformed drift, form a cliff nearly 1 km long and up to 20 m high. Unlike the higher cliffs of eastern Møn, the chalk masses at Hvideklint do not form distinct ridges that can be traced inland.

Undisturbed chalk bedrock underlies southern Møn at elevations of –25 m to –50 m (Ter-Borch and Tychsen 1987). At the surface, exposures display deformed Quaternary strata and detached chalk floes. Hvideklint has been examined by

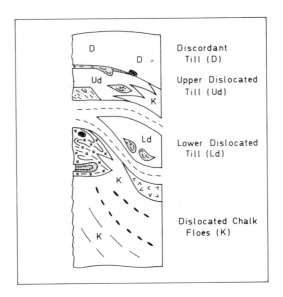

Fig. 5–13. Composite stratigraphic column for Hvideklint. The upper portions of chalk floes are often brecciated (V symbol) or sheared together with drift forming melange (X symbol). Modified from Aber (1979, fig. 4).

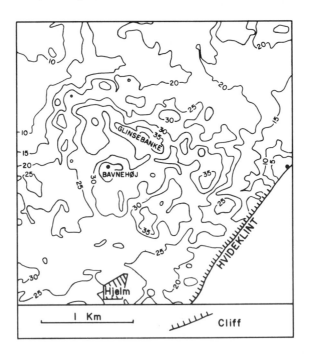

Fig. 5–14. Topographic map of cupola-hill at Hvideklint. Contour interval = 5 m; small closed depressions not shown. Small dot at right edge indicates starting point at fishing hamlet for measured section of Hvideklint (fig. 5–16).

Fig. 5–15. Aerial photograph of Hjelm (lower left) and Hvideklint vicinity, southern Møn. Note rolling topography, which is largely cultivated, and lack of ridge morphology. Scale bar is 500 m long; north toward top. Copyright Geodætisk Institut, Denmark, 1974 (A. 29/88).

several geologists (Haarsted 1956; Berthelsen *et al.* 1977; Aber 1979; Berthelsen 1979).

Although the chalk at Hvideklint is lithologically similar to that of Møns Klint, the Hvideklint chalk is older. The western portion of Hvideklint contains Campanian chalk, and chalk in the eastern portion is from brachiopod zone 2 of the lower Maastrichtian (Surlyk 1971). The Campanian chalk is the oldest chalk exposed in Denmark. Although the chalk floes appear to rest in more-or-less horizontal positions, they are in fact cut by numerous faults and are brecciated in many places.

Three tills along with associated stratified drift are exposed at Hvideklint (fig. 5-13). A discordant till overlies upper and lower dislocated tills. Large bodies of deformed stratified drift are found at or near the base of the upper dislocated till. The lower dislocated till also includes some extremely contorted lenses of stratified

86   CHAPTER 5

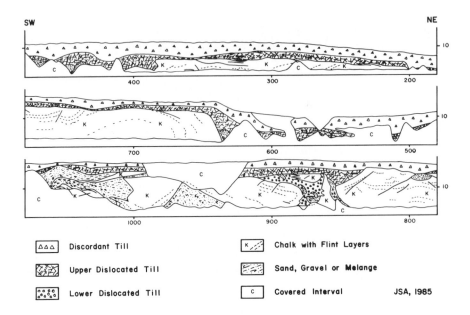

Fig. 5-16. Measured profile of Hvideklint section as it appeared in 1979. Base of section is at beach level. Note 2X vertical exaggeration slightly distorts geometry. Scale in m; adapted from Aber (1979, fig. 3).

drift. The till and stratified drift overlie detached floes of chalk, in which chalk breccia and chalk-till melange are locally developed.

Pebble counts reveal that both the discordant and upper dislocated tills are generally characterized by a large amount of local material and about equal contents of crystalline and Paleozoic types typical of northeastern derivation. However, the lower dislocated till has noticeably less chalk and more Paleozoic limestone indicative of a Baltic source. The presence of reworked Eemian foraminifera in the discordant and upper dislocated tills indicates Weichselian age for those tills (Aber 1979).

The cupola-hill north of Hjelm has an oval outline and contains many irregular small hillocks and closed depressions (fig. 5-14). It rises to a maximum elevation of 44 m at Bavnehøj. Only one subdued ridge, Glinsebanke, is present. Chalk is exposed at several places on Glinsebanke, so it is probably the upturned end of a chalk floe. There are otherwise no obvious linear trends in the cupola-hill morphology; land slopes are gentle to moderate throughout the hill (fig. 5-15).

Hvideklint displays a progressive increase in deformation from a minimum at the northeastern end to a maximum toward the southwest. Thus, the cliff is conveniently divided into four portions (fig. 5-16):

1. Eastern chalk floe (200-410 m) – A slightly deformed chalk raft, tilted toward the southwest, is overlain continuously by the discordant and upper dislocated tills. Small thrusts and brecciated zones are present within the chalk, particularly near its southwestern end.

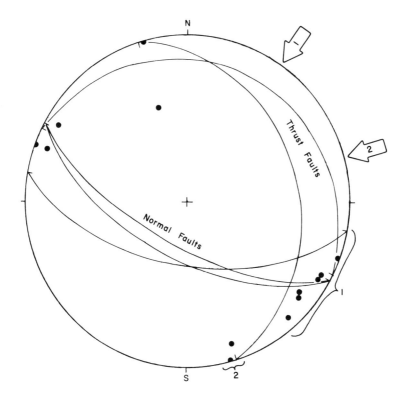

Fig. 5–17. Hvideklint structural data plotted on an equal-area stereonet. Solid dots = fold axes; arcs = faults. Note grouping of data into two clusters.

2. Central drift sequence (410–640 m) – Discordant and upper dislocated tills carry across this portion of Hvideklint. Large bodies of deformed stratified drift are found near the base of the upper dislocated till, and no solid chalk is present.

3. Central chalk floe (640–855 m) – The largest single chalk body at Hvideklint forms a broad anticline in which a conjugate thrust system is developed. The chalk is thrust over isoclinally folded and sheared stratified drift at its southwestern end.

4. Western chalk floes (855–1080 m) – Highly disturbed chalk and drift display extreme stretching and thinning toward the southwest. Penetrative deformation has caused intermingling of chalk and drift in isoclinal folds and shear bands (Plate III). A body of lower dislocated till is almost entirely enclosed by deformed chalk near the eastern end of this portion (860–890 m).

The orientations of glaciotectonic structures are consistent along the cliff section. Fold axes are mostly near horizontal, and all trend either southeast or northwest. Thrust and normal faults also strike northwest-southeast (fig. 5–17). These structural data correspond to northeasterly ice movement at Hvideklint. The orientation data are clustered in two groupings: (1) 100 to 140° and (2) 160 to 165°. This suggests that there were actually two episodes of structural disturbance associated with ice advances from about 30° and 70°. The chalk floes were presumably derived from somewhere in the Fakse Bugt vicinity north of Møn.

Southern Møn was subjected to the same late Weichselian ice advances that affected eastern Møn (Chapter 3). The discordant and upper dislocated tills relate to the Main Weichselian advances. The upper dislocated till was probably deposited by the initial advance from the northeast, and so correlates with the Mid Danish Till. This advance transported the chalk and constructed the ice-shoved hill.

The discordant till was then deposited by the Storebælt readvance from the east-northeast, and so corresponds to the North Sjælland Till. This advance caused some additional structural disturbance and smoothing of the cupola-hill. Hvideklint was apparently little altered by subsequent Young Baltic advances. The lower dislocated till is problematical; it could be the Ristinge Klint Till or it could relate to a still older Saalian advance.

I. Upper portion of Hanklit, Mors, Denmark. Detached and folded mass of Fur Formation claystone is enclosed by gravel and rests on till. Note brittle fracturing in core of fold (left) and plastic stretching in nose of fold (right). Dimensions of view approximately 30 x 50 m; see Fig. 3–2. Photo by J. Aber, 1979; copyright by Hunter Textbooks, Inc. 1988.

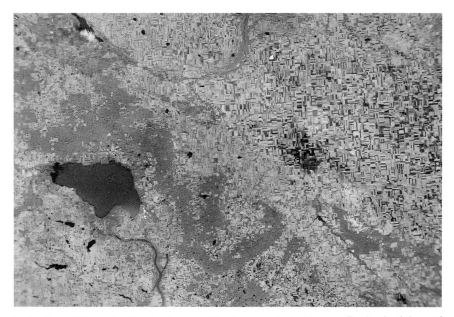

II. Portion of false-color Landsat image of Regina area, Saskatchewan. Regina is pink spot in upper right corner; Dirt Hills and Cactus Hills are reddish-pink loops in lower middle; Old Wives Lake is dark blue to left. Image 21215–16530, 21 May 1978; north toward top; see Fig. 3–8.

III. Chalk-till melange from western portion of Hvideklint, Møn, Denmark. Note penetrative mixing and isoclinal folding of chalk and drift. Shovel app. 75 cm long; see Chap. 5; photo by J. Aber, 1979.

IV. Kansas Drift stratotype at Atchison, Kansas. Brown Upper Kansas Till (uppermost right) rests on Atchison Formation sand, which is underlain by gray Lower Kansas Till (lower left). The gray till forms a large diapir where the person is standing. Ladder is 7 m long; see Fig. 7–5; photo by J. Aber, 1987.

CHAPTER 6

MEGABLOCKS AND RAFTS

Introduction

The common occurrence of flat-lying glacial rafts or megablocks throughout glaciated plains was not appreciated until fairly recently (Stalker 1976; Sauer 1978; Ruszczyńska-Szenajch 1987). These megablocks are more-or-less horizontal, slightly deformed, and are often buried under or within thick drift giving little or no morphologic clue to their presence in the subsurface. They may, in fact, easily be mistaken for bedrock, if deep exposures or drilling logs are not available. In other cases, the megablocks form flat-topped buttes, small plateaus, or irregular hills, which have also been mistaken for bedrock outliers.

Most megablocks are composed of poorly to moderately consolidated Mesozoic or Cenozoic sedimentary strata, but some consist of well-consolidated rocks or unconsolidated Quaternary strata. All megablocks exhibit some signs of deformation – shear zones, folds, faults, brecciation, *etc.* – as a result of ice pushing. This criterion distinguishes megablocks and rafts from large, undeformed erratic blocks (Aber 1985a). The sources of many megablocks are unknown or poorly identified, but most were probably transported only a few km. Source depressions for megablocks cannot usually be identified.

A typical small megablock of upper Pennsylvanian (Paleozoic) limestone is present at the surface near the Kansan glacial limit west of Topeka, Kansas (Map 1; Dellwig and Baldwin 1965). The megablock consists of the Tarkio Limestone Member of the Zeandale Formation. Its exposed dimensions are 50 m north-south by 150 m east-west with a thickness of only 1 to 2 m. The internal structures include fracturing, thrusting and rotation of limestone blocks.

The underlying Willard Shale is normally about 12 m thick in the vicinity. However, below the megablock only thin (30 cm) remnants of brecciated shale and glacial debris are present, resting on a striated surface of the undeformed Elmont Limestone. Striations agree with structural features indicating ice movement from the northwest. This megablock was probably derived from nearby (< 1 km away) by horizontal sliding.

Long-distance transportation of megablocks is documented in some cases. A group of famous rafts at Lukow in east-central Poland (Map 2) are composed of Jurassic clay derived from Lithuania > 300 km to the northeast (Jahn 1950). Some of these rafts are > 20 m thick, and they were probably transported in a frozen condition (Ruszczyńska-Szenajch 1976).

The Cooking Lake megablock, located near Edmonton in central Alberta (Map 1), lies at the surface. It is about 10 m thick and covers an area of roughly 10 km^2.

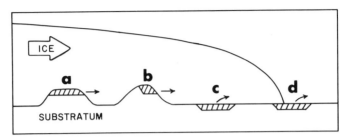

Fig. 6–1. Possible means of detaching megablocks: a = beheading of butte, b = plucking from lee of hill, c = scooping from depression, d = thrusting at ice front. Vertical scale exaggerated; a, b and c may occur anywhere under the ice. Taken from Aber (1985a, fig. 3).

This megablock consists of Lower Cretaceous strata from the Grand Rapids Formation, which has its nearest outcrop in the Thickwood Hills > 300 km to the northeast (Stalker 1976).

All such megablocks have one trait in common: they are remarkably thin (typically < 30 m) compared to their lateral dimensions (often > 1 km^2), although a few are much larger. They are thin rock slices, which could only have been transported by freezing onto the underside of a glacier. Thus during transportation, the megablock was effectively the base of the ice sheet. Deposition occurred when basal melting released the megablock from the ice. At that point, continued glacier

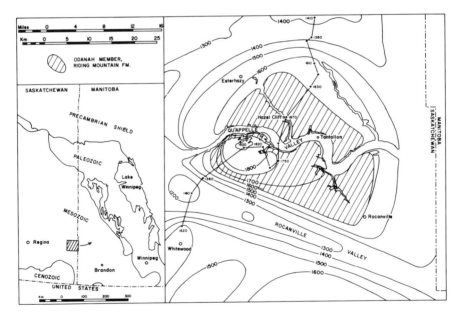

Fig. 6–2. Bedrock contour map of Esterhazy vicinity, Saskatchewan. The 1600-foot contour between Esterhazy and Rocanville defines position of megablock. Small dots show test wells used in constructing cross section (fig. 6–4). Elevations of bedrock surface in test wells shown in feet; contour interval = 100 feet (approx. 30 m). Adapted from Christiansen (1971b) and Sauer (1978, fig. 6).

Fig. 6–3. Aerial photograph of Qu'Appelle Valley, eastern Saskatchewan. Bedrock outcropping along walls of valley is part of huge megablock. Scale bar is 2 km long; north toward top. Tantallon is small town in upper left corner. Aerial photograph A21748–94; copyright 1970. Her Majesty the Queen in Right of Canada, reproduced from the collection of the National Air Photo Library with permission of Energy, Mines and Resources Canada.

movement could have further deformed the megablock, eroded it away, or covered it with till.

The fact that such megablocks are scattered over broad regions behind ice-margin positions supports a subglacial origin for many megablocks (fig. 6–1). Some megablocks may also have been initially pushed in proglacial settings (Ruszczyńska-Szenajch 1976). In certain cases, megablocks are aligned in chains of hills decreasing in size away from their sources.

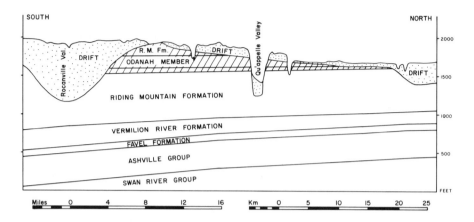

Fig. 6–4. Subsurface cross section showing megablock (diagonal lining) of Riding Mountain Formation at Esterhazy. See Fig. 6–2 for location; vertical exaggeration = 42 X. Adapted from Christiansen (1971b).

The following case examples span a great size range from the huge megablock (about 1000 km^2) at Esterhazy, Saskatchewan, to medium megablocks (1–10 km^2) of southern Alberta, to a chain of small rafts (< 1 km^2 each) at Kvarnby, Sweden. Although diverse in their other characteristics, these examples do share one feature in common; they all consist of Cretaceous strata.

Esterhazy, Saskatchewan, Canada

During the course of geological and ground-water reconnaissance mapping in eastern Saskatchewan, a truly enormous megablock of Cretaceous shale was discovered at Esterhazy by Christiansen (1971b). The megablock covers an area roughly 1000 km^2 in extent (fig. 6–2). The region is part of the Saskatchewan Plain, a drift-mantled plain of relatively low relief underlain by soft Cretaceous sedimentary bedrock. The megablock has no morphologic expression at the surface.

The monotonous plain is broken by a large spillway channel, the Qu'Appelle Valley, which cuts through the middle of the megablock (fig. 6–3). The Qu'Appelle Valley is > 100 m deep and about 2 km wide. Its bottom is drift filled, and bedrock outcrops are found along its walls. The position of the Qu'Appelle Valley is in no way related to the presence of the megablock. Cross cutting of the megablock by the valley was merely a fortuitous coincidence.

Bedrock in eastern Saskatchewan is generally undeformed, dipping gently toward the south (fig. 6–4). The Cretaceous formations are mostly clastic strata consisting mainly of siltstone, claystone and shale. Hard, siliceous shale of the Odanah Member, Riding Mountain Formation is more consolidated than other strata and forms the main mass of the megablock.

Highly folded and faulted bedrock of the Riding Mountain Formation is found west of Hazel Cliff, east of Tantallon, and south of the Qu'Appelle Valley both in surface exposures and in drill holes. Breccia, slickensides, and mylonite are common microstructures. This deformation is confined to bedrock above about 1500 to 1600 feet elevation (Christiansen 1971b). The 1600-foot contour may be used as a minimum outline for the megablock.

The general plan of the megablock is an egg-shaped oval, roughly 38 km long and 30 km wide. Bedrock in the west-central portion reaches 1920 feet elevation, indicating a maximum thickness of about 100 m. The megablock is, thus, at least 300 times wider and 380 times longer than it is thick. Assuming an average thickness of 60 m and an area of 1000 km^2, the megablock's volume is estimated to be 60 km^3.

A test well drilled southwest of Esterhazy near the western end of the megablock intersected 2 m of till after penetrating 80 m of brecciated and mylonitic bedrock (Christiansen 1971b). Similarly disturbed bedrock rests directly on undeformed bedrock south of the Qu'Appelle Valley. The actual decollement zone is located in Riding Mountain Formation claystone below the Odanah Member, and some claystone above the Odanah Member is also part of the megablock.

The megablock is situated on the northern edge of a major buried glacial valley, the Rocanville Valley. This valley is filled with up to 60 m of sand capped by till. Melt-water erosion and filling of the valley presumably took place during an earlier (pre-late Wisconsin) glaciation, and the till cap was deposited during the late Wisconsin glaciation. It appears that the Rocanville Valley truncates the southern side of the megablock, and thus, emplacement of the megablock must predate cutting of the Rocanville Valley.

Ice flow in this part of Saskatchewan was generally from the northeast; however, neither the direction of emplacement nor the source of the megablock are known. The megablock may not have moved far, perhaps less lateral displacement than its own width, in order to produce the observed structures. The only conceivable means of displacing a megablock of such size was by freezing onto the bottom of an overriding ice sheet, in which case the megablock became the basal layer of the ice sheet. It is highly improbable that this megablock could have been pushed in front of an advancing glacier. Subglacial sliding of permafrozen material over a thawed substratum seems to be the most likely explanation for displacement of the Esterhazy megablock.

Southern Alberta, Canada

Numerous, large megablocks comprised mainly of Cretaceous strata are scattered throughout the plains of southern Alberta (fig. 6–5). The total number of such megablocks is unknown, but is undoubtedly great. Most known megablocks are partly or wholly buried beneath thick drift, and some are known only from drilling

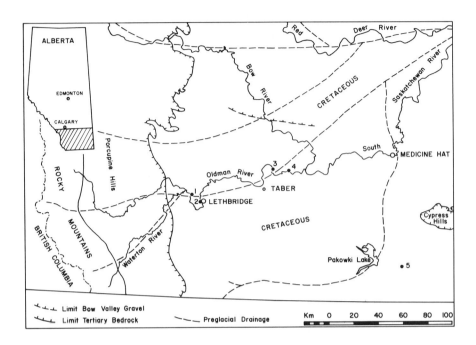

Fig. 6–5. Selected megablocks of southern Alberta: 1 = Kipp, 2 = Laundry Hill, 3 = Driftwood Bend, 4 = Wolf Island, 5 = Catchem. Adapted from Stalker (1976, fig. 35.1); preglacial drainage from Prest (1976, fig. XII–9).

records. Where megablocks make up a substantial volume of drift-plain deposits, as in many parts of southern Alberta, the surface landform may be called a *megablock plain*.

Stalker (1973b, 1976) was apparently the first geologist to recognize that many large rock masses previously thought to be bedrock or interglacial deposits are in fact glacial rafts, to which he applied the term megablocks. These megablocks are underlain by Quaternary till, gravel, or silt, and they may extend for several km, but are always quite thin (< 30 m). Selected megablocks are described in the following paragraphs. Many of these megablocks are exposed in bluffs where modern valleys cross buried valleys, but megablocks are also known in other situations.

Kipp – Bluffs along the eastern side of the Oldman River, about 15 km west of Lethbridge, contain a 1.5-km-long megablock about 25 m above river level and 60 m below the bluff top. Tills are present both above and below the megablock. Preglacial gravel resting on Bearpaw Formation shale is present at the base of the bluff. The megablock includes a melange of sandstone with clay partings, coaly seams and ironstone bands. The source of this megablock is unknown; likewise its direction and distance of transportation are unknown.

Laundry Hill – This megablock is exposed in bluffs along the eastern side of the Oldman River and in the coulee (ravine) near Laundry Hill entirely within the city

Fig. 6-6. Map of western Baltic region showing location of chalk bodies at Kvarnby and outcrops of upper Cretaceous bedrock (black). Taken from Ringberg *et al.* (1984, fig. 1).

limits of Lethbridge. The megablock contains up to 14 m of dark, bentonitic shale and white, poorly indurated sandstone. It is overlain by multiple tills, gravel, silts, and varved lake beds. At least two tills are present below the megablock as well. Megablock thickness averages about 5 m, and it was originally > 1 km wide. The megablock is nearly horizontal, and in many parts it displays remarkably little internal disturbance. Its source and distance of movement are uncertain.

Driftwood Bend – One of the largest megablocks yet discovered in Alberta outcrops in the bluff along the eastern side of the Oldman River about 15 km northeast of Taber. It consists of interbedded Cretaceous shale, coal, and sandstone strata. Intermittent exposures extend for > 3 km along the bluff, and seismic shothole logs indicate the megablock extends a considerable distance behind the bluff. Maximum thickness is about 25 m with average thickness being about 10 m. Stalker (1976) estimated the Driftwood Bend megablock covers at least 10 km^2.

The continuity and intact nature of this megablock have been stressed, but the megablock may actually include several discrete blocks lying adjacent to each other at about the same level. Some blocks are nearly flat-lying, and others are strongly deformed. Deformation is minimal toward the southern end of the section and increases northward. At least one block near the northern end forms a large recumbent fold that is stretched southward into melange. The megablock is underlain by some 25 m of drift including till, and is overlain by 15 m of similar material. Immediately beneath the megablock, shear planes and slickensides are developed in the till. Once again, the source and distance of travel of this megablock are not known.

Wolf Island – A megablock composed of Cretaceous shale, coal and bentonitic sandstone is exposed for a distance of nearly 1.5 km along the northern bluff of the Oldman River about 20 km east-northeast of Taber. It is both underlain and overlain by tills and other drift in the 75-m-high section. The megablock is more-or-less horizontal, up to 13 m thick, and is locally deformed particularly near its base, where it intermingles with underlying till.

The megablock also includes up to 3 m of preglacial gravel resting on the Cretaceous strata. The gravel has a Bow Valley lithology, which is not normally found in the Oldman River system. The southern limit of Bow Valley gravel is more than 50 km to the north (fig. 6–5).

Catchem – This megablock is known from only two drill holes about 20 km east of Pakowki Lake. The megablock is probably no more than 1 km^2 in extent, and its source and distance of transportation are not known.

Megablocks are seemingly ubiquitous in the Alberta Plain. The source and direction of movement for many are unknown, but some did travel considerable distance. The megablocks are not related to any single glaciation or particular topographic setting. Their emplacement appears to be widely scattered in time and space.

Individual megablocks were probably transported by freezing onto the bottom of an overriding ice sheet. Thin or discontinuous permafrost would have facilitated the detachment of large, thin megablocks and their incorporation onto the base of the ice sheet. Basal melting would have later allowed separation of megablocks from the ice. At that point, megablocks could have been partly disrupted, pulled apart, folded or sheared together with drift.

Kvarnby, Skåne, Sweden

A series of small upper Cretaceous chalk rafts, known since the beginning of this century (Holst 1903, 1911), is situated immediately east of Malmö in Skåne, southwestern Sweden (fig. 6–6). These chalk bodies were recently investigated by Ringberg (1980, 1983) in connection with geological mapping in the Malmö area, and comprehensive analysis was carried out by Ringberg et al. (1984). The chalk rafts are aligned in a northeast-trending chain stretching 4.5 km long in a zone 700 to 800 m wide (fig. 6–7). These chalk rafts are exposed in quarries, and several other chalk bodies are present near the surface or buried along this trend (fig. 6–8).

The chain of chalk bodies is situated in a drift-mantled region along a mor-

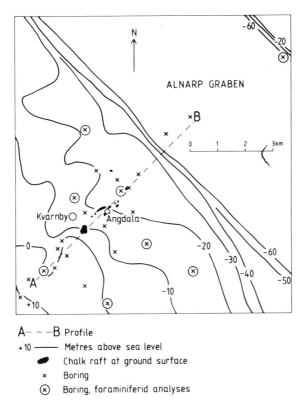

Fig. 6–7. Bedrock contour map of Kvarnby vicinity showing locations of chalk rafts and borings. Position of cross section (fig. 6–8) indicated; contours on bedrock surface in m above or below sea level. From Ringberg *et al.* (1984, fig. 3).

phologic boundary. A hummocky moraine of small hills is developed to the northwest at elevations 10–20 m above sea level. To the east and southeast, the land is 15–60 m above sea level and more rugged. Kvarnby is located on the southwestern margin of a deeply buried bedrock valley, the Alnarp Graben, although the position of the chalk rafts is probably not related to the presence of the buried trough.

It is known from numerous test wells that Danian (lower Paleocene) limestone, 50–100 m thick, forms the bedrock surface under this entire region. Soft, white Maastrichtian chalk (upper Cretaceous) underlies the Danian limestone. Microfossils in chalk rafts at Kvarnby indicate late, although not latest, Maastrichtian age. Outcrops of similar chalk bedrock are found to the northeast within the highly faulted Fennoscandian Border Zone at Romeleåsen, to the south and west on Rügen, Møn and at Stevns, and on the intervening Baltic sea floor (fig. 6–6).

The two largest chalk rafts, which reach a maximum thickness of 25 to 30 m, are exposed in quarries at Kvarnby and Ängdala. The two exposures display close similarities (fig. 6–9). The Kvarnby Till underlies and is dislocated between the

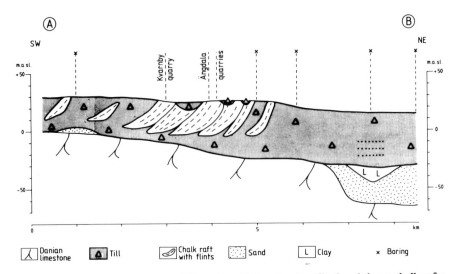

Fig. 6–8. Schematic cross section of Kvarnby vicinity showing tilted and thrust chalk rafts. Large vertical exaggeration; taken from Ringberg *et al.* (1984, fig. 4).

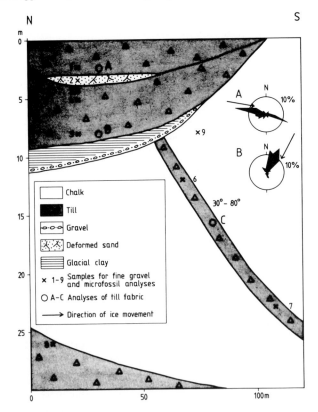

Fig. 6–9. Composite stratigraphy of chalk quarries at Kvarnby and Ängdala. A = Sunnanå Till, B = S. Sallerup Till, C = Kvarnby Till. Taken from Ringberg *et al.* (1984, fig. 5).

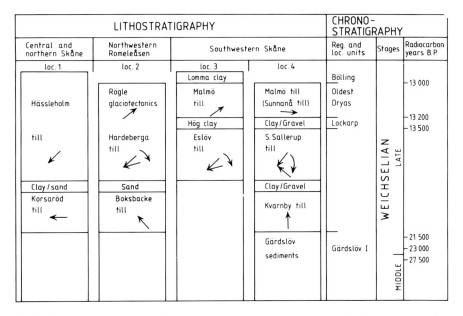

Fig. 6–10. Schematic diagram of till stratigraphy, ice movement, and chalk emplacement at Kvarnby (loc. 4) and possible correlations to other locations in southern Sweden. Taken from Ringberg (1988, fig. 10).

chalk rafts. The chalk bodies are included stratigraphically within the Kvarnby Till, which is up to 30–50 m thick and rests on Danian limestone (Ringberg 1988). The chalk is thoroughly brecciated into pebble- and cobble-sized fragments; flint is also fractured in some places, but still retains continuous bedding.

Chalk rafts, flint beds, and interlayered till all dip consistently 30° to 80° toward the south or south-southwest. The till is rich in Paleozoic limestone and other Baltic indicators along with reworked Eemian, Tertiary, Jurassic and Paleozoic microfossils. All evidence points to a Baltic source for the Kvarnby Till and included chalk rafts, with ice movement coming from the south.

The S. Sallerup Till is up to 6 m thick and rests discordantly on the chalk rafts with up to 1 m of clay and gravel locally present between chalk and till. Small chalk floes within the till dip toward the northeast. The lower portion of the till is dominated by local material and crystalline erratics derived from the northeast and has a northeasterly till fabric. Higher in the till, northeast indicators decrease and Baltic material increases again, reflecting a shift in ice movement direction during till deposition.

The Sunnanå Till (= Malmö Till of Ringberg 1988) is separated from the S. Sallerup Till by a thin layer of deformed sand thought to be about 13,300 years BP. This youngest till is rich in Baltic rock types, much like the Kvarnby Till. Striations and till fabrics show ice movement from southwesterly to westerly directions (Ringberg 1988). Presumably it was deposited by a Baltic ice lobe that advanced into the Öresund strait from the south and then spread eastward into Skåne. However, the exact nature of this so-called Low Baltic advance in southern Sweden is the subject of some debate (Lagerlund 1987; Ringberg 1988).

All three tills at Kvarnby and Ängdala are believed to be late Weichselian in age (fig. 6–10). The first and last ice advances were by ice lobes following the Baltic depression to the south and west of Skåne, whereas the middle advance was the main Weichselian glaciation. This main advance apparently shifted its direction of movement over Skåne from northeasterly, to easterly, to southeasterly while continuously covering the region.

The chalk rafts were emplaced by the initial Baltic advance from the south, and were subsequently modified by the main advance from the northeast. The chalk rafts were carried more than 25 km from the Baltic sea floor to Kvarnby, probably while frozen to the base of the glacier. The chalk was likely transported as one or a few larger rafts, which were pulled apart during deposition and thrust together with the Kvarnby Till forming the northeast-trending chain of rafts.

CHAPTER 7

DIAPIRS, INTRUSIONS AND WEDGES

Introduction

Various kinds of diapirs, intrusions and wedges are widely and commonly reported from glaciogenic sequences. Intrusive structures range in lateral size from only a few cm to > 100 m (fig. 1–4) and may have vertical dimensions of several 10s of m. Intrusive structures rarely give rise to distinctive or noticeable landforms, but they may dominate the subsurface character of a glacial sequence (fig. 7–1).

Intrusions include all those structures in which one material was injected or squeezed in a mobile state into the body of another material. Some mixing between injected and host materials may take place along the intrusion boundaries, but the two materials still remain distinct. In other words, the injected and host materials do not become homogenized. The intrusive material is most usually a clay- or silt-rich sediment; whereas the host material may be almost any kind of unconsolidated sediment or soft bedrock.

Creation of intrusive structures implies a significant difference in physical

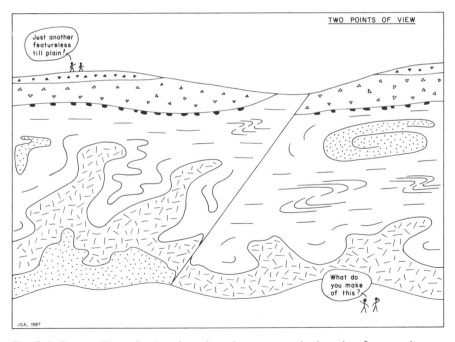

Fig. 7–1. Cartoon illustrating how large intrusive structures in the subsurface may have no morphologic expression at the surface.

103

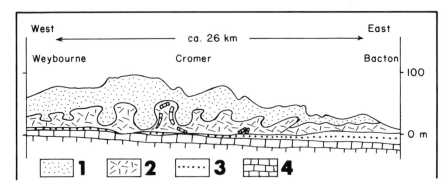

Fig. 7–2. Schematic section between Weybourne and Bacton, Norfolk showing large diapirs of clay-rich till containing chalk floes. 1 = Gimmingham Sands and Briton's Lanes Gravels, 2 = Cromer Tills containing chalk floes, 3 = Cromer Forest Series Bed, and 4 = Senonian chalk bedrock. Large vertical exaggeration; adapted from Banham (1975, fig. 14).

properties of the materials at the time of intrusion. The injected fine-grained sediment behaved as a fluid, for example quick-clay, that was mobilized by pore water trapped under high pressure. The host material, conversely, behaved in a less ductile or even brittle manner. It is generally agreed that such intrusions took place in a subglacial, water-saturated condition with intergranular movement as the main mechanism of deformation (Brodzikowski and van Loon 1985).

A distinction should be drawn at this point between those intrusions that originate from below and from above the host material. Diapirs, stocks, plugs, dikes and sills result when mobilized sediment intrudes from below the host material. Such structures resemble shallow intrusions of low-viscosity (basaltic) magma or diapirs formed by salt flowage.

The force compelling upward intrusion is simply gravity acting on low-density material, such as clay- or silt-rich sediment, buried below higher-density, coarse-grained sediment or lithified strata. Diapirs may develop when fine-grained strata are compacted and become mobilized by high-pressure water. The sediment then behaves as a fluid and seeks to rise or escape toward zones of lower pressure. Upward intrusion continues as long as the sediment remains fluid until a density equilibrium is achieved or until the source of intrusive sediment is depleted.

The most important question concerns the nature of the compacting load that mobilized intrusion. Only those cases where glacier ice provided the diapir-inducing load should be considered glaciotectonic. Diapirs and other intrusions created by soft-sediment deformation induced solely by deposition of thick overburden are not truly glaciotectonic, even if they are found in glaciogenic sequences.

The classic cliff exposures of chalk floes within contorted drift at Cromer, Norfolk, United Kingdom (Map 2) illustrate the importance of identifying the loading agent. This famous section was long thought comparable to chalk cliffs of Møn, Denmark (Slater 1926). However, Banham (1975) demonstrated that large, mushroom-shaped diapirs of clay-rich Cromer Tills were created by loading

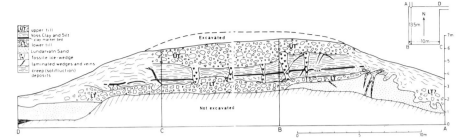

Fig. 7–3. Section showing stratigraphy and wedge structures in walls of a building excavation (A-D) near Voss, Norway. Taken from Mangerud and Skreden (1972, fig. 6). Wedges and veins later reinterpreted as glaciotectonic structures.

beneath thick outwash sand and gravel, not by glacier advance (fig. 7–2). The chalk floes were incorporated in the First Cromer Till during an earlier and unrelated glaciation. Thus, the major diapiric structures of the contorted drift are not glaciotectonic, according to Banham (1988b). This non-glaciotectonic interpretation remains controversial, however (Hart 1988).

Wedges, veins, and fissures commonly penetrate from above along cracks opened in a brittle host material. Common wedge sediment varies from homogeneous till to laminated sand, silt or clay. The positions of wedges range from nearly vertical to nearly horizontal. Their strike orientations may be related to direction of ice movement or to direction of subglacial pressure gradient. Of course, wedge positions may also be controlled by zones of weakness within the host material.

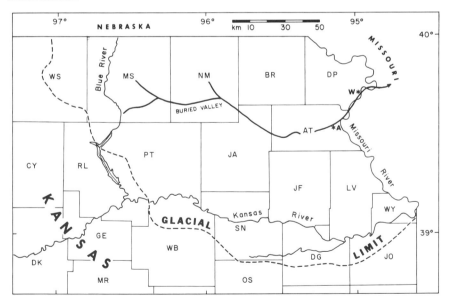

Fig. 7–4. Northeastern Kansas showing glacial limit, buried valleys (Dreeszen and Burchett 1971, fig. 1), and locations of described sites: A = Atchison, W = Wathena. AT = Atchison County, DP = Doniphan County.

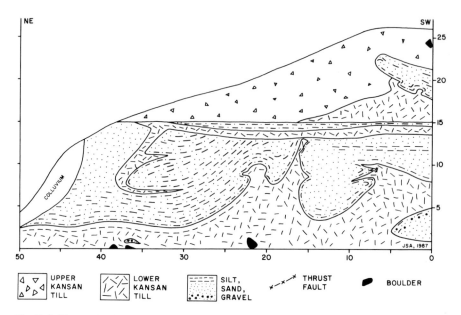

Fig. 7–5. Kansas Drift stratotype at Atchison, Kansas as it appeared in 1987. Lower Kansas Till (below) intrudes up into Atchison Formation sand (middle), which is overlain by Upper Kansas Till (top). Scale in m; see Plate IV.

A strong tendency exists among geologists to interpret wedge structures as fossil ice wedges, thermal crack fillings, or other permafrost features, perhaps because their genesis is so well understood (Washburn 1980). At Voss, Norway (Map 2) for example, unsorted till wedges and laminated clay and silt veins were originally considered to be permafrost features (fig. 7–3). However, these wedges were later reinterpreted as glaciotectonic structures (Mangerud *et al.* 1981).

Glaciotectonic wedges form in dynamic situations, whereas permafrost wedge fillings are created in a passive manner. Slickensides, foliation, grooves, mixing zones, drag folds, small thrusts, wedge apophyses and host xenoliths are all features that indicate a dynamic, intrusive origin for wedges. Lack of such structures is evidence for a passive wedge genesis, possibly related to permafrost conditions. Some wedges may have a dual genesis: originally formed in permafrost and later deformed glaciotectonically.

Diapir and wedge structures that are glaciotectonic in origin are presented in the following case examples: Kansas Drift, Kansas; Herdla Moraines, Norway; and Systofte, Denmark. In these cases, the intrusive sediment is clay or silt rich and was injected as a result of glacier overriding. None of these structures has any morphologic expression.

Kansas Drift, Kansas, United States

The Kansan glaciation was the most extensive ice coverage to take place on the

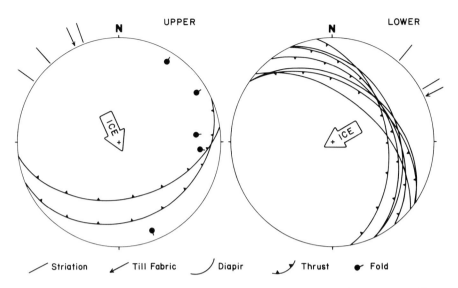

Fig. 7–6. Summary equal-area stereonets of structural and directional features at Atchison section. Lower Kansas Till was disturbed by northeasterly ice movement; northwesterly ice advance was responsible for the Upper Kansas Till.

Great Plains of central North America and is recognized as an important early glaciation of the Pleistocene. This glaciation is represented by the Kansas Drift, a lithostratigraphic unit defined in northeastern Kansas (Aber 1985b, 1988d).

The Kansas Drift is quite old, roughly 600,000 to 700,000 years BP (Aber *et al.* 1988), and is probably the oldest regionally preserved Pleistocene glacial sequence on land. Because of its great age, original glacial morphology is gone, and the upper portion of the Kansas Drift is greatly altered by weathering and erosion. The Kansan glaciation created many, scattered glaciotectonic deformations both in drift and in consolidated Paleozoic limestone and shale bedrock of the region (Dellwig and Baldwin 1965).

Where thick Kansas Drift fills preglacial valleys, as at Atchison and Wathena (fig. 7–4), its original stratification and structures can be seen in deep exposures. The Kansas Drift stratotype at Atchison includes three formations (Plate IV):

1. Upper Kansas Till – brown, compact, stony, clay-rich till containing irregular blocks of stratified drift (Atchison Fm.), 3–15 m thick, northwesterly fabric and striations.
2. Atchison Formation – rippled, fine to very fine silty sand with scattered lenses of pebbly sand, basal portion mainly sandy pebble gravel, thickness 10 to 20 m.
3. Lower Kansas Till – gray, compact, stony, clay-rich till containing wood fragments, 1 to 12 m exposed, northeasterly fabric and striations.

In the stratotype section, the Lower Kansas Till is dislocated in a pair of large

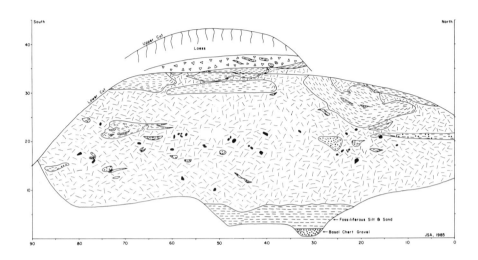

Fig. 7–7. Wathena section as it appeared in 1985. Scale in m; symbols same as Fig. 7–5.

diapirs which intrude up into the overlying sand (fig. 7–5). The larger diapir in the center of the section has been exposed for at least half a century. Dellwig and Baldwin (1965) interpreted the structure as a thrust anticline formed by frictional drag of ice readvancing over frozen sediments. The ice came from an easterly to northeasterly direction. Stream erosion during recent years has enlarged the section, revealing the second diapir in the lower part plus additional structures higher up.

The two diapirs are strikingly similar in form and orientation. Both have enlarged, asymmetrical heads; both terminate upward at about the same level; and both dip northeastward. Till fabrics, striations on boulders, thrust faults, and the diapirs themselves show consistent orientations, measured over a period of many years, indicating ice pushing from about 60° (fig. 7–6). The intrusive character of the diapirs indicates deformation in a thawed and water-saturated state.

The Upper Kansas Till caps the section and contains deformed masses of Atchison Formation sand. Ice movement, which laid down the upper till, also produced a pair of recumbent folds in the underlying sand near the northeastern end of the exposure. The axes of these two folds plunge slightly toward the northeast. The recumbent folds reach down to the level of the diapir heads.

The peculiar shapes of the diapir heads are explained by low-angle faults or overturned folds on their left sides. These secondary structural features are oriented toward the east-northeast. Directional features within the Upper Kansas Till and deformations in the subjacent sand, including diapir heads, all indicate ice movement from the northwest (fig. 7–6).

The Wathena site (fig. 7–4) is located in gravel pits along the Missouri valley bluff. These pits have been worked intermittently for many years, resulting in a large exposure (fig. 7–7). Pre-Kansan basal chert gravels are overlain by horizontally bedded fine sand and silt, above which gray Lower Kansas Till comprises most of the exposure. Bodies of very fine silty sand of the Atchison Formation toward the top are overlain by Upper Kansas Till and younger loess.

In several places, the lower till has intruded upward in the sand forming irregular plugs and sills. Likewise, certain bodies of Atchison Formation sand appear to have sunk into the lower till. These foundered sand bodies contain small blobs or lenses of gray till. Like the till diapirs at Atchison, the till intrusions here must have occurred in a mobilized state, when high fluid pressure had reduced the shear strength of the till's clay matrix and the intruded sand was not frozen.

The evidence from Atchison and Wathena is consistent with ice movement from the northeast during deposition and subsequent displacement of the Lower Kansas Till and Atchison Formation. Multiple northeasterly advances are probable. The initial advance laid down the lower till, and sand was then conformably deposited over the till during a brief ice recession. The sand accumulated in proglacial lakes within valleys that were blocked by ice to the northeast. Owing to its lake-bottom position, the Atchison Formation probably did not become permafrozen.

Renewed northeasterly advances then dislocated the lake-floor till and sand with thrusting and diapiric intrusions. After retreat of the northeastern ice, a northwesterly advance deformed the upper portion of the Atchison Formation, including diapir heads, and deposited the Upper Kansas Till.

Loading of Lower Kansas Till by either Atchison Formation sand or ice readvance could be responsible for intrusion of diapirs. However, soft-sediment deformation seems unlikely in this case, because the maximum thickness of the Atchison Formation is only 30 m. Assuming this maximum thickness plus an additional 30 m of lake water, an increased load of only about 9 kg/cm^2 was imposed on the underlying till. This is not a great load, equivalent to roughly 100 m of glacier ice.

The Lower Kansas Till displays the compact, well-oriented, parallel fabric typical of basal tills laid down beneath active glacier ice. Ice thickness during deposition of the Lower Kansas Till cannot be stated with certainty, but it surely exceeded 100 m. Therefore, deposition of Atchison Formation sand in proglacial lakes was insufficient to cause soft-sediment deformation of the previously compacted Lower Kansas Till. Subsequent ice overriding during renewed advances could have caused sufficient loading to mobilize the till. Rapid ice movement during a surge is the most attractive mechanism, as strong pressure gradients would have developed within the overridden material.

The Atchison Formation sand served as a permeable medium into which mobilized clay-till diapirs could intrude. Such diapirs do not occur in upland sites, where the Atchison Formation is lacking. At least four and possibly as many as seven glacial advances are documented from upland sites between Wathena and Atchison (Dort 1966, 1985). The multiple upland tills are, perhaps, the results of repeated advances, which created diapirs in thicker glaciolacustrine sediments of the valleys.

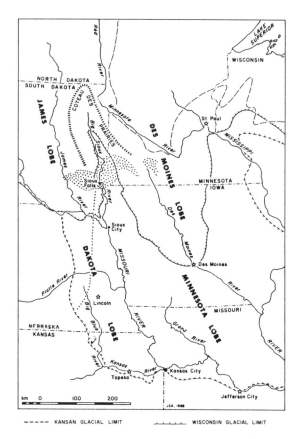

Fig. 7–8. Map of the northern Great Plains showing glacial features. James and Des Moines lobes are Wisconsin; Dakota and Minnesota lobes are Kansan. Dotted zone is Sioux Quartzite bedrock ridge at southern end of Coteau des Prairies. Adapted from Aber (1988g, fig. 22).

The stratigraphy and structures displayed at the Kansas Drift stratotype demonstrate the development of two ice lobes during the Kansan glaciation (fig. 7–8). The Minnesota lobe advanced southward through Iowa, into Missouri, and entered Kansas from the northeast. Conversely, the Dakota lobe came from the Dakotas, across Nebraska, and moved into Kansas from the northwest. These two lobes followed broad bedrock troughs either side of the Coteau des Prairies upland and Sioux Quartzite bedrock ridge in southwestern Minnesota. At times the two lobes coalesced over the Coteau forming a broad Kansan ice fan to the south. The Kansan ice lobes and fan were evidently quite dynamic.

Herdla Moraines, Norway

The Herdla Moraines is a morphostratigraphic unit defined by Aarseth and

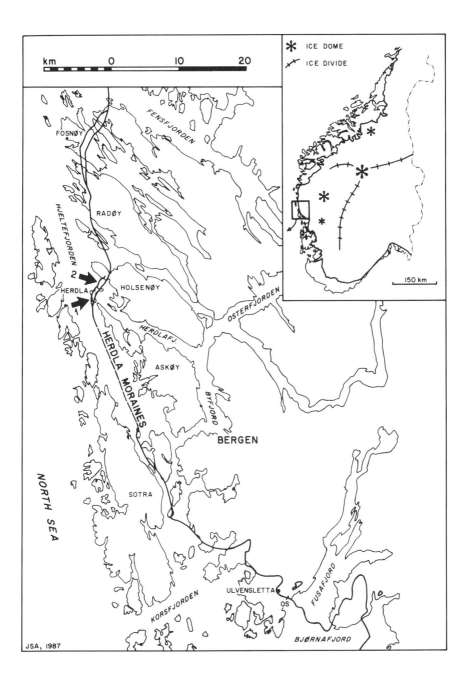

Fig. 7–9. Location map for the Herdla Moraines western Norway: 1 = Herdla island (fig. 7–10); 2 = Herdlaflaket. Inset map shows Younger Dryas ice margin and reconstruction of ice domes and divides. Taken from Aber and Aarseth (1988, fig. 1).

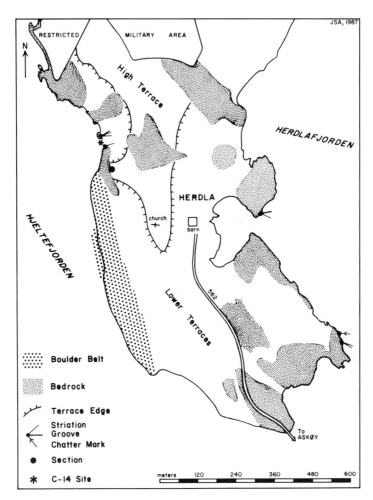

Fig. 7–10. Map of southern Herdla island showing geologic and glacial features. Taken from Aber and Aarseth (1988, fig. 2).

Mangerud (1974). It forms one segment of the moraine system that was deposited around the periphery of Norway (fig. 7–9) during the Younger Dryas glaciation (latest Weichselian). Moraine sediments are thickest and morphologically most prominent below the contemporaneous marine limit. Large ice-contact submarine fans and deltas, up to 100 m thick, are preserved in valleys and fjords both above and below present sea level. The Younger Dryas moraines of western Norway are disturbed in many places by small glaciotectonic structures, including diapirs, as a result of oscillations by the active ice margin (Sollid and Reite 1983).

The moraine on Herdla island (fig. 7–9) is morphologically the most conspicuous in the Bergen region. The island lies across the western end of Herdlafjord and includes several bedrock knobs, between and around which the glaciomarine

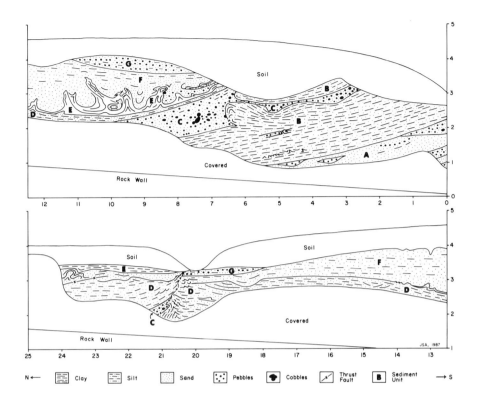

Fig. 7–11. Measured section on western Herdla island (see fig. 7–10 for location) showing internal structure of moraine deposits. Base of section is approximately 18 m above sea level; scale in m. Taken from Aber and Aarseth (1988, fig. 4).

deposits are found (fig. 7–10). Till is present on the proximal (eastern) side of the island, and a boulder belt along the western edge marks the outer limit reached by the Younger Dryas ice sheet. Most striations and grooves on bedrock trend west or southwest between 240° and 275°. A smaller, younger group trends toward the northwest between 290° and 300°.

Clay- and silt-rich layers within the sediment are strongly consolidated. In connection with seismic stratigraphy over Herdlaflaket (fig. 7–9), the acoustic velocity of Herdla Moraines sediment was estimated to be 1800 m/s (Aber and Aarseth 1988). This compares with 1600 m/s for unconsolidated sediment. This degree of consolidation could only be caused by compaction beneath overriding ice, because the sediments have never been deeply buried otherwise.

Internal structure of the moraine is exposed in a road-cut section on the western side of the island above a bedrock knob. Moraine sediment there comprises several units (A-F, fig. 7–11) of interbedded clay, silt, sand and gravel deposits. This sediment was presumably laid down in water some 10–15 m deep in front of the Younger Dryas ice sheet; a rapidly fluctuating sedimentary environment is indicated. The sediments were deposited as part of a submarine fan, when the ice margin was located along the eastern edge of Herdla island.

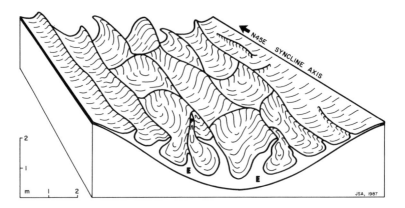

Fig. 7–12. Cut-away block diagram showing idealized geometry of unit E diapirs within the broad syncline at Herdla section. Diapirs form a network of bent partitions. Taken from Aber and Aarseth (1988, fig. 6).

The sediments are disturbed by several secondary structures, the largest of which is a broad syncline, whose axis plunges gently toward the northeast (fig. 7–11). In the southern limb of the syncline, units B and C display structures typical of soft-sediment deformation. Coarse gravel of unit C appears to have settled or collapsed into unit B silt and sand. This deformation happened before deposition of unit D and folding of the syncline.

Toward the center of the syncline, unit E forms a series of small, but elegant diapirs. The biggest diapirs, up to 1 m high, occupy the syncline trough, and progressively smaller diapirs and overturned folds are found on the syncline flanks. The smaller flank diapirs are overturned inward (toward the trough), whereas bigger trough diapirs bend outward. Sand of unit F above the diapirs is foliated and wraps smoothly around each diapir. Individual diapirs are thin, only 10–20 cm thick, and have curved shapes in the third dimension.

The diapirs appear to be thin-walled partitions forming a bent, three-dimensional network or grid of intrusions (fig. 7–12). The partition network is bent according to its position within the broad syncline. Clayey silt sediment forming the diapirs thickens toward the center of the syncline, as if the silt had flowed toward the fold trough during intrusion of the diapirs. The overall geometry of the diapir network implies that the diapirs were created at the same time the syncline was folded. Flowage and intrusion of unit E sediment caused slight thickening in the trough portion of the syncline.

The northern portion of the section is disrupted by thrust faults and associated folds. Unit C is thrown up along a northwest-dipping thrust that cuts through the entire sequence. This fault and related folds are subparallel with the broad syncline axis and presumably formed at the same time as the syncline. Together they produce overall structural shortening of the section in a SE-NW direction. Another

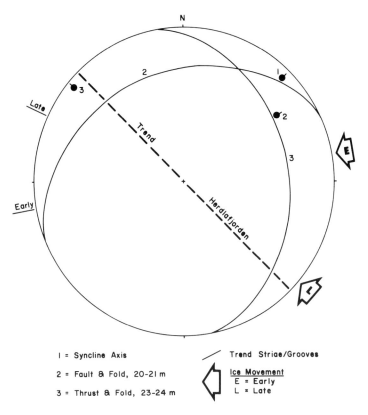

Fig. 7–13. Summary equal-area stereonet plot of directional data for various features of the Herdla vicinity. Early phase of ice movement from east-northeast; late phase from southeast subparallel to Herdlafjord. Taken from Aber and Aarseth (1988, fig. 9).

thrust fault with overturned drag folds can be seen at the far northern end of the section entirely within unit D. This fault strikes/dips 350/30° NE. These features indicate fault displacement in an ENE-WSW direction.

At Herdla, two phases of moraine development took place corresponding to local changes in movement of the Younger Dryas ice sheet (fig. 7–13). During the early phase, the ice margin reached eastern Herdla, and a submarine fan was constructed. Ice movement at this time came from the east-northeast, and slight shifting of the ice margin created minor thrusting of moraine sediment.

The direction of ice movement subsequently changed to southeasterly and the ice margin advanced to the western edge of Herdla. Most of the deformation in the section on western Herdla, including folding and thrusting of the broad syncline, occurred during this late phase. Consolidation of the sediment and intrusion of diapirs in a fluidized state also took place at this time, due to increased loading by overriding ice. This direction of ice movement was essentially parallel to Herdlafjord.

116   CHAPTER 7

The overall position of Herdla Moraines and pattern of older striations are related to radial outflow from an ice dome to the northeast of Bergen during the early phase (fig. 7–9). The ice sheet must have been fairly thick, as ice flow was largely independent of local topography. The shift to fjord-parallel ice movement in the Herdla vicinity probably occurred as the ice sheet was becoming thinner, and ice flow was thus more responsive to local topography.

The large Younger Dryas readvance in western Norway also caused crustal depression with resulting marine transgression (Anundsen 1985). Higher sea level may have increased the rate of calving and thus accelerated drawdown of the ice sheet. Another possible consequence of the marine transgression was the occurrence of glacier surges along the fjords. Surging is a common form of glacial advance for similar fjord glaciers in Spitsbergen today (Elverhøi *et al.* 1983).

Rapid loading of saturated sediments during a glacier surge is an ideal setting for creation of diapirs and other glaciotectonic features. The glaciotectonic structures present on Herdla are, thus, related largely to activity of the local fjord glacier and may not correlate with glaciotectonic features in other portions of the Herdla Moraines.

Systofte, Falster, Denmark

Humlum (1978) described a large till wedge exposed in a gravel pit near Systofte, on the island of Falster, southeastern Denmark (Map 2). Various till intrusions have

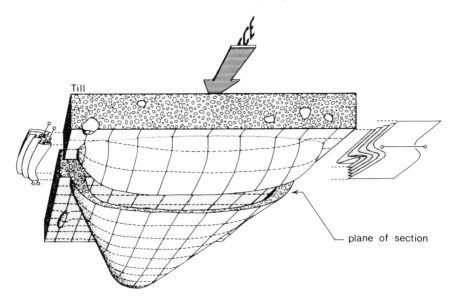

Fig. 7–14. Block diagram showing 3-dimensional geometry (from below) of till wedge at Systofte, Denmark. Small drag folds adjacent to wedge shown to left; large recumbent folds below till cap to right. Section about 12 m across; taken from Humlum (1978, fig. 6).

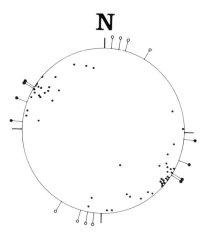

Fig. 7–15. Equal-angular stereonet of till fabric (small dots) and folds at Systofte, Denmark. Solid fold axes are small drag folds; open-circle axes are large recumbent folds. Taken from Humlum (1978, fig. 7).

been noted in many parts of Denmark before (Hansen 1930; Berthelsen 1974) and must be rather common. Most of these appear to have originated from tills buried below the host. However, the wedge at Systofte was injected from above.

The Systofte wedge is located in stratified sand and gravel of a kame that was overridden and is covered with about 2 m of till. The kame forms an elongated cupola-hill trending north-south and standing approximately 10 m above the surrounding landscape. Cross bedding within the kame indicates southward or southwestward melt-water flow when the stratified drift was laid down.

The wedge takes the form of a shallow trough, about 12 m across, that cuts at least 2.5 m deep into underlying sand and gravel (fig. 7–14). The wedge is about 1 m thick near its top, where it is continuous with the overlying till cap, but thins to only 10 cm toward the bottom. The upper part of the wedge cuts across stratification in the sand and gravel, and numerous small drag folds are present adjacent to the wedge. Axes of these drag folds trend between 95° and 120°, subparallel to the axis of the wedge trough. The deeper part of the wedge becomes parallel with bedding in adjacent stratified drift.

The outer 1–2 cm of the wedge displays slight mixing of sand derived from adjacent walls, whereas the center of the wedge is enriched in fine silt and clay particles relative to the till cap. This textural zonation reflects the sorting and mixing action of fluid flow during intrusion of the wedge. The till cap has a well-developed fabric that indicates ice movement toward approximately 305° (fig. 7–15). The axes of large recumbent folds below the till cap are roughly at right angles to this direction. Taken together, the structural and fabric information are consistent with ice movement toward the northwest during intrusion of the till wedge and deposition of the till cap.

The last Weichselian ice to cover southeastern Denmark was the Young Baltic glaciation, consisting of East-Jylland and Bælthav advances. These advances came from the southeast on Falster (Houmark-Nielsen 1981, fig. 12). Humlum (1978) concluded that both till and stratified drift were unfrozen at the time of wedge intrusion. Sorting and mixing of sediment within the wedge and intensive drag folding of adjacent sand could hardly happen in frozen material. This implies that the Systofte wedge was intruded beneath a temperate glacier base and that the Young Baltic glaciation took place over unfrozen ground at this location.

CHAPTER 8

APPLIED GLACIOTECTONICS

Introduction

Glaciotectonic deformation produces locally severe disruption of the normal geology and terrain, and therefore its recognition and structure are important for many human activities. The principal deleterious results of glaciotectonism are: (1) disruption and deformation of the stratigraphic sequence, (2) reduction of sediment or rock strength, and (3) major increase in sediment variability. Activities, for which recognition of glacially thrust terrain is important, include mine planning and operation, drift prospecting and mineral exploration, construction, and soils mapping and utilization.

Glaciotectonic terrain may be expected to be texturally, lithologically, geochemically and geotechnically anomalous compared to the surrounding terrain. The terrain may provide data on the local subsurface stratigraphy, because thrust masses may include sediment and bedrock shoved up from the substratum. The depression upglacier from a thrust mass, if present, may serve as a window into the underlying stratigraphy. Ice-thrust terrain will often have thinner drift cover than surrounding terrain, and in some places the bedrock itself may be exposed. Folding may produce inversion of stratigraphy that if unrecognized would cause problems during interpretation of drilling data.

Problems in soils mapping over glaciotectonic terrain originate from the high lateral variability of the deformed substratum. The parent material, which is of major importance to soils mapping, may change from sandstone, to shale, to till, to clay over a few 10s of m. If bedrock or other subsurface sediments contain harmful substances, thrusting can bring these materials into the pedologic zone. For example, bedrock thrust to the surface can increase soil salinity problems because of high salt content of the bedrock compared to the normal till cover.

One aspect of glaciotectonism often overlooked in applied studies is the depression left at the source of thrusting. This could be important during mineral-exploration and mine-planning phases of resource development (Fenton and Pawlowicz, in prep.). The false assumption that an ore body continues across a glacial depression would yield overly optimistic estimates of total reserves. Sudden discovery of the absence of ore during mining could cause disruption of the mining schedule with potentially serious economic implications. Use of a combination of test drilling and surface geophysics between test holes is the most economical way to avoid a surprise discovery of this kind.

In some coal mines of western Canada, glaciotectonism has locally removed the coal from areas up to 1 km^2. At one mine, ice thrusting extended deep enough to

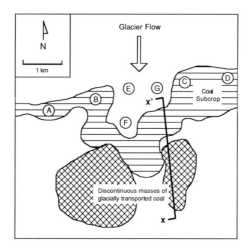

Fig. 8–1. Map showing subcrop of coal in relation to ice thrust depression and transported coal.

remove about 12 million m³ of coal leaving a depression that is partially masked by the cover of later Quaternary sediment (figs. 8–1 and 8–2). With test holes drilled only at positions A, B, C and D, the interpretation would be that the coal subcrop is continuous. Additional drilling, such as holes E, F and G, would indicate that coal is missing. The transported coal remaining downglacier from the depression is of no use for mining, because the coal exists in small hills close to the surface. This position means the coal has been oxidized, so that its calorific value is greatly reduced and the moisture content increased.

The following examples are situations where glaciotectonic disruption of normal bedrock stratigraphy and structure has created serious hazards or economic consequences for human activities. The examples are from large- and small-scale,

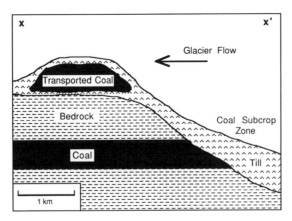

Fig. 8–2. Schematic cross section showing the depression formed by glaciotectonism and downglacier hill including some of the coal transported from the depression.

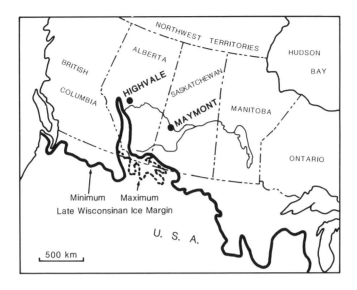

Fig. 8–3. Location of Alberta and Saskatchewan examples with postulated maximum and minimum late Wisconsin glacial limits (after Prest 1984).

open-pit mining operations, respectively in Alberta, Canada and Fur, Denmark, and from highway construction through small-composite ridges in Saskatchewan, Canada. They illustrate the importance of recognizing the special substratum conditions in glaciotectonic terrain.

Highwall Failure, Highvale Coal Mine, Alberta, Canada

Highwall failures in glacially thrust bedrock have occurred in several coal mines of western Canada, including the Highvale Mine west of Edmonton, Alberta (fig. 8–3). The mine is owned by TransAlta Utilities Corporation; production is 11.4 million tons per year; and the adjacent Sundance Power Plant is the largest electrical generating station in western Canada (Tapics 1984).

Open-pit mining here involves two basic operations: (1) removal of the overburden above the coal using a dragline and (2) removal of the coal seams using power shovels (fig. 8–4). The dragline operates from a bench above the coal, removing the overburden above and below the bench and casting it onto the spoil pile beyond the current mining pit.

Glacially thrust bedrock is weaker than undeformed bedrock, and as a result highwalls and benches cut into this material have a greater tendency to fail. The implications of highwall failure are serious; temporary benches excavated in the highwall serve as transportation corridors and as working surfaces for heavy equipment. In addition to the risks posed to men and equipment, such failures also

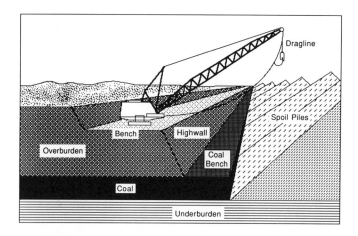

Fig. 8–4. Schematic cross section of typical open pit mine showing overburden bench and highwall.

result in potentially sizable extra costs for rehandling overburden material, disruption of mining schedules, and outright loss of mineable coal.

The regional bedrock geology consists of flat-lying, upper Cretaceous and Paleocene, non-marine, coal-bearing strata. The coal belongs to the Ardley Coal Zone of the Scollard Member of the Paskapoo Formation (Carrigy 1970; Irish 1970; Holter *et al.* 1975). Six distinct and laterally continuous seams are mined.

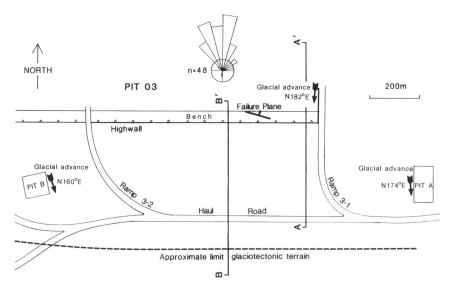

Fig. 8–5. Map of southeastern part of Pit 03 showing areas of glacially deformed sediment, sites of structural observations, and locations of geologic and hydrogeologic cross sections. Modified from Moell *et al.* (1985, fig. 8–2).

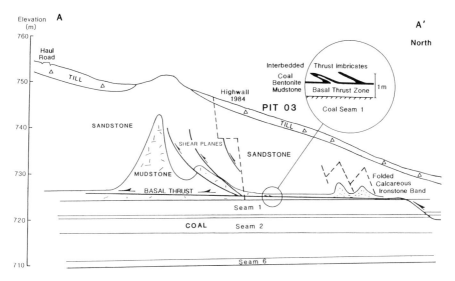

Fig. 8–6. Geologic cross section A-A' (fig. 8–5) showing deformation. Modified from Moell et al. (1985, fig. 8–1).

They are separated by shale and bentonite partings, with a cumulative coal thickness of about 10 m. Quaternary strata consist of a discontinuous cover of glaciofluvial and glaciolacustrine sediment over till (Andriashek et al. 1979). Glaciotectonism of the bedrock is extensive, although most disturbed sites are < 1 km^2 in area.

During 1983 and 1984, a series of highwall failures occurred in Pit 03 between ramps 3–1 and 3–2 (fig. 8–5). As a result, research was initiated to determine the cause and mechanism of the failures. Geologic data were collected through outcrop study, air-photo interpretation, surface geophysical methods, rotary and auger coring, and downhole geophysical logging. Ground-water data were obtained by setting and monitoring a number of piezometer nests. Much of the following information is summarized from Fenton et al. (1983, 1985, 1986) and Moell et al. (1985).

The site stratigraphy is from the top down: till, sandstone, mudstone, interlaminated mudstone, coal and bentonite, and a six-seam coal unit. The till contains abundant clasts and lenses of sheared bedrock material. The underlying sandstone has been subjected to major folding and faulting, although the generally massive nature of this unit commonly prevents recognition of small deformation structures.

The mudstone unit is massive and highly fractured with many of the fractures having polished or slickensided surfaces. This unit is also cut by numerous small shear zones. Where exposed in the highwall, each is about 1 mm thick, clay filled, concave upward, and is about 1 m^2 in areal extent. The contact with the overlying sandstone was in most places, when freshly excavated, the site of ground-water discharge in both 1983 and 1984. A mass of thrust bedrock overlies till in the western third of the area (fig. 8–5).

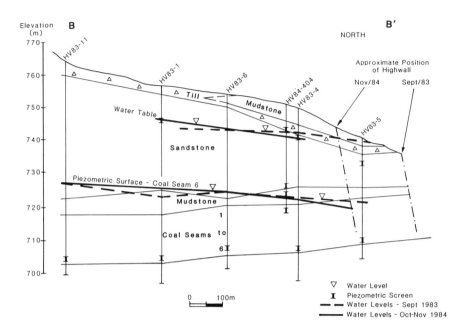

Fig. 8–7. Hydrogeologic cross section B-B' (fig. 8–5) showing piezometric surface of coal seam 6 and water table in the glacially deformed overburden. Modified from Moell et al. (1985, fig. 7–6).

The glaciers advanced generally from the north, and the bedrock has been folded, faulted, and crushed by compression and thrusting. The thrust and shear planes produced during deformation dip generally northward (fig. 8–6). Detailed data on the larger glaciotectonic structures were collected from the three locations where sections are transverse to the structural grain of the area. Averaging the data from each site indicates displacement directions of 174° for pit A, 182° for ramp 03, and 160° for pit B (fig. 8–5). These directions agree with the average of 170° obtained from measurements of smaller shear planes exposed in the fractured mudstone unit.

The base of disturbance is a shear zone that rises stratigraphically southward from a position that involves deformation of seams 1 and 2 to a position above seam 1 near the haul road (fig. 8–6). The shear plane is believed to die out at some point south of the haul road. Where the shear plane immediately overlies seam 1, movement was along a 1-m-thick zone of interlaminated mudstone, bentonite and coal and is well illustrated by small folds in coal laminae.

Hydrogeologic data indicate the glacially deformed overburden is almost completely saturated with ground water and that ground-water flow is directed from south to north toward the highwall (fig. 8–7). However, stratigraphic relationships act to inhibit drainage of this saturated, poorly consolidated bedrock. Since 1983, ground-water levels have declined only slightly, leaving 20 m of saturated

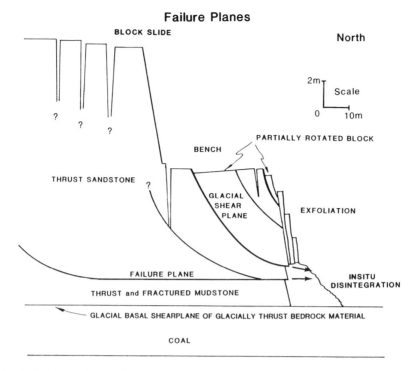

Fig. 8–8. Schematic drawing showing observational data and model for highwall failure. Modified from Moell et al. (1985, fig. 8–6).

overburden. Consequently, the overburden remains poorly drained, and there is no evidence of improvement in drainage conditions.

Surface and subsurface data indicate that highwall failure was composite in nature with four types contributing to the overall result. These four are: (1) block sliding, (2) block rotation, (3) exfoliation or spalling, and (4) *in-situ* distintegration of large failed blocks (fig. 8–8). Block sliding resulted in a series of vertical fractures marked by cracks, up to 4 m deep. These fractures trended approximately parallel to the highwall face. They developed over the area between ramps 3–1 and 3–2 and extended into the land at least 50 m behind (south of) the upper edge of the highwall.

The rotational slumps varied in size, but were generally 50 to 100 m long, roughly semicircular in plan, with direction of rotation approximately northward. One slump, which was observed within 12 hours of failure, appeared to have moved along a pre-existing surface, likely a glacial shear plane. The remaining two failure types were volumetrically less significant. Exfoliation consisted of falling sheets or slices of the sandstone about 1 m to 3 m thick. *In-situ* disintegration was the gradual disaggregation or reduction in size of sandstone blocks that had accumulated at the base of the slope as a result of the previous types of failure.

The failure cycle begins as the dragline starts a new cut at the eastern end of the

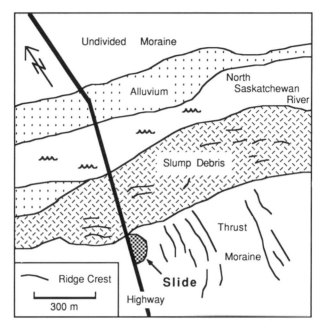

Fig. 8–9. Surficial geologic map of Maymount area, Saskatchewan. Adapted from Krahn *et al.* (1979, fig. 3).

pit and moves westward excavating the overburden bench as it moves. With time, the initial excavation site starts to undergo the above types of failures. These progress along the cut both in the direction of excavation and southward behind the cut, until within a few months the entire bench has been destroyed. At any particular site exfoliation occurs first followed by rotational slumping and block sliding. The rotational slumps, particularly the large ones, are likely the result of remobilization of pre-existing glacial shear planes (fig. 8–8).

The block failures are believed to be caused by elevated pore-water pressures near the contact of the sandstone and fractured mudstone units. The situation was compounded by the saturated and weakened condition of the overlying sandstone. The dip of the sandstone/mudstone contact toward the pit undoubtedly also contributed to block sliding. Vertical fractures are a manifestion of the block sliding and may be the result of the sliding itself, but are more likely due to opening of pre-existing joints or are perhaps a combination of joints and shear planes. The base of failure, near the sandstone-mudstone contact, lies above the basal shear zone of the ice-thrust mass (fig. 8–6) demonstrating that failure is taking place within the thrust mass.

In summary, overall highwall failure was the product of both the geologic and hydrogeologic regime at the site. Failure resulted from excavation into bedrock that had been crushed, folded, and faulted by glaciotectonism and that was situated in an area of ground-water saturation and elevated pore-water pressure. Orientation of

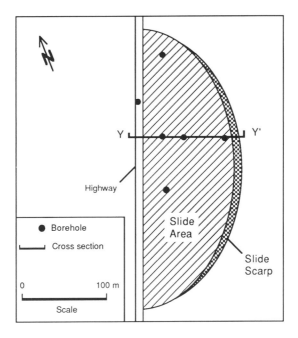

Fig. 8–10. Map of slide showing location of cross section (fig. 8–11) and test holes. Modified from Krahn et al. (1979, fig. 7).

the highwall also contributed to failure by exposing glacially induced shear planes that dipped into the pit. Hydrologic conditions were caused by minimal hydraulic conductivity together with the northward hydraulic gradient into the mine. Foreknowledge about location, structure, dimensions, and hydrology of glaciotectonically disrupted terrain is important to both mine planning and management.

Highway Construction, Maymount, Saskatchewan, Canada

Recognition, structure, and composition of glaciotectonic terrain is important for geotechnical site investigation for highways, bridges, dams, and related constructions. The following example is from the Maymount area of Saskatchewan (fig. 8–3). Construction began in 1973 on a highway and bridge to cross the North Saskatchewan River valley. When highway excavation was nearing completion, a massive failure occurred. Subsequent investigation indicated the failure was due to low shear strength of the glacially deformed bedrock. This bedrock also contributed to difficulties in construction of the bridge abutment and to extensive natural landslides adjacent to the river (Sauer 1978; Krahn et al. 1979).

Surficial geology of the region (fig. 8–9) consists of glacially deformed bedrock and landslide debris on the southern side of the river and of flood-plain sediment

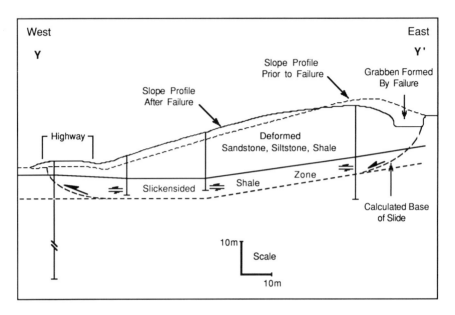

Fig. 8–11. East-west cross section through Maymount slide. Modified from Krahn *et al.* (1979, figs. 8 and 12).

and thick undivided moraine sediment on the northern side. Bedrock on the southern side is interbedded sandstone and shale of the upper Cretaceous Judith River Formation. The Judith River Formation is folded, faulted, brecciated, and slickensided. Also in places, inclusions of till were found below the formation. Sauer (1978) concluded that the entire area underlain by the Judith River Formation on the southern side of the valley is likely a huge thrust mass.

The slide occurred in the eastern highway embankment near the northern margin of the glaciotectonic terrain (fig. 8–9) and has been described in detail by Krahn *et al.* (1979). The slide occurred between 6 p.m. and the next morning and consisted of a block about 270 m long by 120 m wide that moved westward, down slope, toward the highway. Following the slide, an east-west line of test holes was drilled and tube samples were collected (fig. 8–10).

Test-hole data revealed the site geology to be interbedded sandstone, siltstone and shale overlying a zone of highly brecciated and slickensided shale (fig. 8–11). The slickensided zone on the eastern side of the road cut dips westward, but orientation of this zone on the western side of the cut is unknown. Glaciotectonism had deformed the bedrock to a depth of 45 m. Large folds were revealed in the upper 3 m in a scarp created by the failure. Piezometers set during the drilling indicated the site was a recharge area of downward gradient with ground-water levels close to the upper boundary of the slickensided zone.

Geotechnical testing of tube samples demonstrated that shear strength of shale composing the slickensided zone had been reduced from peak to residual values. This was interpreted to be the direct result of glaciotectonism of the bedrock. The

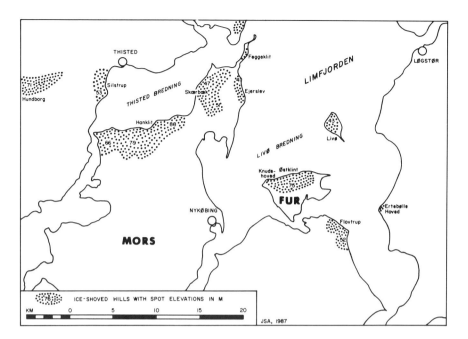

Fig. 8–12. Ice-shoved hills of western Limfjord region. Hills built largely or partly of disturbed moler (Fur Formation). Map based on interpretation of topographic maps plus aerial photographs.

slip surface for failure was assumed to lie within the slickensided zone, because there was no conclusive evidence uncovered to determine its exact location except near the crest and toe of the failure.

Factors contributing to the failure were low strength of the disturbed bedrock and dip of the slickensided zone toward the excavation. Stability of the embankment west of the highway was concluded to be either because the slickensided zone did not dip toward the road or because negative ground water pore-pressure conditions contributed to greater sediment strength.

Diatomite Quarries, Fur, Denmark

A distinctive, perhaps unique, bedrock formation is found in the western Limfjord district of northern Jylland, Denmark. The material, locally known as *moler* (mo-clay), consists of clayey diatomite with a large number of volcanic ashes (Pedersen and Surlyk 1983). Moler was highly susceptible to glaciotectonic disturbance; in all surface exposures moler is dislocated, folded and faulted. A variety of hill-hole pairs, large and small composite-ridges, and cupola-hills are found on islands of the Limfjord and on the surrounding mainland (fig. 8–12). These hills are composed largely or partly of deformed moler along with glacial sediments.

The moler strata are formally designated as the Fur Formation (Pedersen and

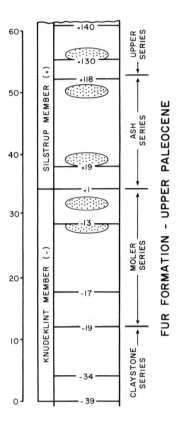

Fig. 8–13. Composite stratigraphic column for the Fur Formation showing positions of some prominent ash beds (numbered) and concretion zones (stipled). Based on Pedersen and Surlyk (1983, fig. 13).

Surlyk 1983) with age established as latest Paleocene (fig. 8–13). Diatomite of the Fur Formation is highly porous with a low density (0.8 g/cc) consisting of marine diatoms and clay minerals (mainly montmorillonite). Dark-gray to black ash layers are for the most part basaltic tephras of glass and mineral particles. The ash layers are conformably interbedded with diatomite and contrast sharply with the light-colored diatomite layers. Individual ash layers are continuous and nearly constant in thickness throughout the Fur Formation, allowing for a detailed tephra chronology.

The unusual properties of moler have several industrial applications, so it has been quarried for many decades at various locations, principally on northern Mors and northern Fur. Primary use is for ceramic products. Structural disturbance of moler is the main concern for planning and developing quarries. Quarries generally follow the strike of folded strata until those strata terminate against a fault or the overburden becomes too thick.

Gry (1965) subdivided the formation on the basis of gross lithology and industrial usage into four informal series. From top to bottom, these are (fig. 8–13):

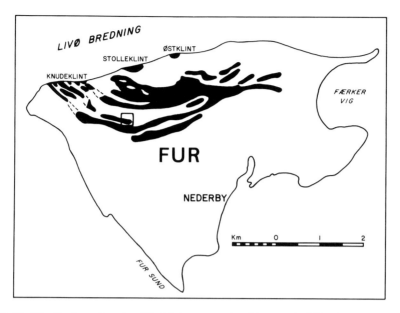

Fig. 8–14. Distribution of moler (black) in composite-ridges and cliffs of northern Fur. Arcuate pattern indicates ice pushing from the north. Study site at Manhøj quarry shown by square; adapted from Pedersen and Petersen (1985, fig. 1).

1. Upper series – almost ash-free diatomite with thin ash layers, +119 to +140, minor industrial use, about 8 m.
2. Ash series – thin diatomite layers interbedded with many, thick ash layers, +1 to +118, no industrial use, about 17 m.
3. Moler series – predominantly diatomite with a few thin ash layers, –19 to –1, major industrial use, about 20 m.
4. Claystone series – dark, clay-rich diatomite with locally silicified beds, ash layers –39 to –20, no industrial use, > 12 m.

The classic study of glaciotectonic disturbance of moler was published by Gry (1940, 1979), who applied methods of structural geology for analyzing the ice-push deformations. Major structures are large, rootless folds of moler with or without deformed glacial strata. Disturbed moler is present in small composite-ridges that make up the northern third of Fur. The ridges reach a maximum height of 75 m and define a gentle arc, concave northward (fig. 8–14). Close agreement exists between structural and topographic trends along the arc, which suggests the composite-ridges were deformed by a northerly ice-lobe advance coming from Livø Bredning.

Difficulty was encountered in excavation at the Manhøj quarry (fig. 8–14), during the early 1980s, as a result of irregular pockets of sand and gravel included within the moler. This sand and gravel contaminates the moler and diminishes its quality as a raw material, which has significance for ultimate extraction of the estimated moler reserve and for day-to-day mining economics. The Geological Survey of Denmark was commissioned to investigate the problem and make recommendations to the quarry operators (Pedersen and Petersen 1985).

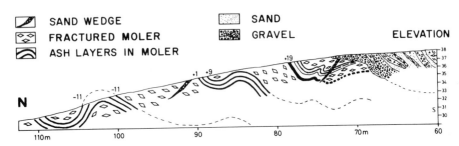

Fig. 8–15. Portion of north-south, test-trench wall about 60 m west of Manhøj quarry. Note north-dipping sand wedges and gravel-filled channel. Adapted from Pedersen (1986, fig. 3).

The middle portion of the Fur Formation, totaling about 35 m in thickness from just above ash +19 to the upper part of the claystone series, is exposed in the quarry. The portion between ash layers –13 to –19 is the quarry interval (see fig. 8–13). Normal thickness of the quarry interval is increased at this location as a result of reverse faulting. The overall structure consists of folds trending NW-SE with a culmination near the center of the quarry. Fold axes plunge about 20° SE in the eastern part of the quarry, whereas folds plunge about 15° NW in the northwestern part. This corresponds closely to the original land topography.

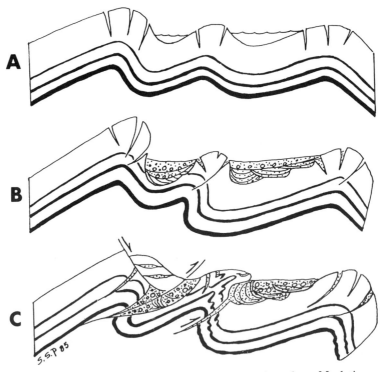

Fig. 8–16. Sequential development of sand wedge structures in moler at Manhøj quarry. A – initial phase of folding, fissure opening and melt-water channels; B – continued folding, fissure development and channel filling; C – final thrust faulting and collapse of anticlines. Adapted from Pedersen and Petersen (1988, fig. 4).

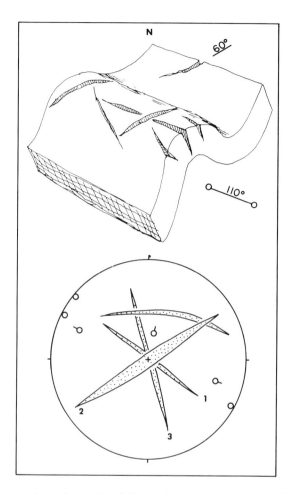

Fig. 8–17. Above: schematic model of fissure development on crest of moler anticline. Below: stereographic projection of fold axes (open circles) and fissures (stipled). Fissure set 1 at 110° represents tension fractures parallel to fold axis; sets 2 and 3 at 60° and 165° are here interpreted as shear fractures. Adapted from Pedersen and Petersen (1985, fig. 11).

Glaciofluvial sediment, consisting of sand and pebble- to cobble-sized gravel, fills channels cut into the folded moler and in places contains reworked masses of brecciated moler. The channels follow syncline troughs and are cut down nearly to ash layer –11. Sand and gravel also form steeply inclined wedges that extend down to 10 m depth within the moler (fig. 8–15). The wedges are up to 30 cm wide near their tops and become narrower downward to only a few cm width before pinching out into irregular fingers. Internal zonation of well-preserved wedges resembles that of ice wedges formed in connection with permafrost.

The development of glaciofluvial channels and wedges was intimately related to folding of the moler during glaciation (fig. 8–16). During the initial stage, anticlines and synclines were folded in moler in front of advancing ice. Folding took place above a decollement zone located at or near the base of the claystone series.

Melt-water channels followed syncline troughs in the glacial foreland, and fissures opened on anticline crests.

Anticlines became overturned with continued deformation, and sediment filling of the open fissures with glaciofluvial sand and gravel took place. Thrusts developed along anticline axial planes and cut up into channel fills. The crests of anticlines collapsed along normal faults into the channels. In this manner, fissures were filled with sediment and were extended or reoriented by continued faulting and folding. Ultimately ice advance overran the area, truncated the folds, and deposited a till cover.

Close relationship is evident between fold orientation and positions of wedge structures (fig. 8–17). The wedges fall in two dominant sets: (1) approximately 110–290°, parallel to fold axes, and (2) roughly 60–240°, oblique to the folds. Set 1 represents tension fractures formed on anticline crests, essentially normal to the direction of maximum compression (= local ice movement) from about 20°.

Pedersen and Petersen (1988) interpreted set 2 wedges as extension fractures; however, true extension fractures should be oriented parallel to maximum compression (Hobbs *et al.* 1976, fig. 7.31). Set 2 wedges are oriented 40° from maximum compression, an angle typical of shear fractures. A third, lesser wedge set is found at about 165–345°. This is 35° from maximum compression and represents conjugate shear fractures in relation to set 2. Finally, there are some wedges oriented in various other directions, which Pedersen and Petersen (1988) thought were inherited from a permafrost polygonal pattern.

This information should be useful in planning future quarry operations. First of all, development of sand wedges is largely a superficial phenomenon that extends no deeper than 10 m into the moler. The distribution of wedges is closely controlled by fold structures, and so zones of sand and gravel contamination within the quarry interval may be predicted and avoided. Careful preliminary studies involving many test pits will be necessary to properly evaluate quality and quantity of moler reserves in ice-shoved terrain of the western Limfjord district.

CHAPTER 9

DISTRIBUTION OF GLACIOTECTONIC PHENOMENA

Continent-Scale Distribution

Glaciotectonic features were considered unusual or rare glacial phenomena when first recognized more than a century ago. This point of view persisted for a long time, because the documented examples of glaciotectonic landforms and structures were few and far between. Recognition and description of glaciotectonic features became increasingly common beginning in the 1950s, leading Sauer (1978) to conclude that such features are probably widespread phenomena in the outer portions of glaciated regions.

Our knowledge concerning the central regions of ice-sheet glaciations has developed more slowly. Fewer investigations of glacial geology were carried out, simply because many geologists assumed that, aside from striations and scattered erratics, not much of interest could be found (Goldthwait 1971).

This point of view must be modified in one important respect. In the zone beneath the ice divide at the centers of former ice sheets, significant drift containing various typical glaciotectonic features is preserved. Thus, a general model for the continent-scale distribution of glaciotectonic phenomena includes three primary zones (Aber and Lundqvist 1988):

1. *Inner zone* in which widespread, small- and moderate-sized glaciotectonic features are developed in older drift.
2. *Intermediate zone* where small, isolated glaciotectonic features are found mainly in locally thick drift of the last glaciation.
3. *Outer zone* in which all manner of large and small glaciotectonic phenomena is present in drift and soft sedimentary bedrock both onshore and offshore.

The zonal model is of course highly generalized; development of each zone may vary according to geological circumstances. The boundaries between zones are sharp in some cases and transitional in other areas. Individual zones are not always fully developed or continuous around the whole area of glaciation. The three zones represent the cumulative results of multiple ice-sheet glaciations during the Pleistocene. The model is concerned with the general or overall distribution of glaciotectonic phenomena, not their local presence or absence; factors related to local distribution will be addressed in the following sections.

The three-zone model can be demonstrated best for the last major glaciation: Laurentide Ice Sheet (Wisconsin) in North America (fig. 9–1; Map 1) and Fennoscandian Ice Sheet (Weichselian) in northern Europe (fig. 9–2; Map 2). The model also applies to earlier glaciations, but glaciotectonic features related to earlier

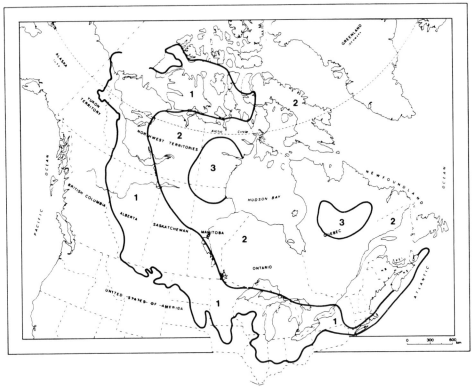

Fig. 9–1. Glacial landscape zones of the Laurentide Ice Sheet: 1 = outer zone of extensive end moraines, hummocky moraines and ice-shoved hills; 2 = intermediate zone with long eskers and ice-flow lineaments; 3 = inner zone of extensive ribbed (Rogen) moraine, drumlins, flutings and eskers. Taken from Dyke and Prest (1987a, fig. 3); copyright Les Presses de l'Université de Montréal.

glaciations are preserved mainly in the outer zone, generally beyond the limits of Wisconsin/Weichselian glaciation. Older glaciotectonic features may also be found in the inner zone of glaciation.

The outer glaciotectonic zone in North America includes the Atlantic Coastal Plain and adjacent continental shelf of southern New England and the Great Plains of the mid-continent beginning in Iowa and stretching westward to the Rocky Mountains and northward to the Arctic continental shelf. Many of the case-history examples in previous chapters, including ones from Massachusetts, North Dakota, Manitoba, Saskatchewan, Alberta, and the Yukon, were drawn from this outer zone.

The outer zone is underlain by soft, poorly consolidated Mesozoic and Cenozoic sedimentary strata consisting mainly of Cretaceous or Tertiary bedrock of clastic composition. Thick and nearly continuous drift covers the bedrock. Large looped end moraines, drumlin fields, older drift, and multiple till sequences are common. All kinds of large and small glaciotectonic phenomena are abundant, and many

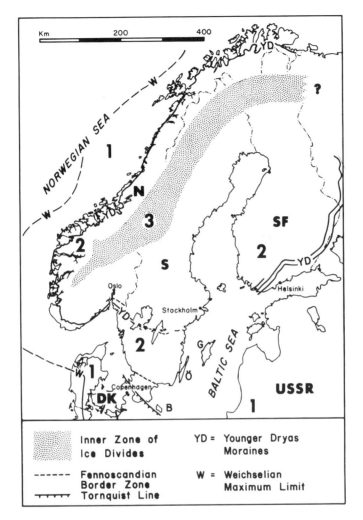

Fig. 9–2. Map of Fennoscandia showing some Weichselian glacial features. Moderately thick and continuous drift containing many glaciotectonic features is present in the inner zone. Numbered zones same as Fig. 9–1; Baltic islands are: B = Bornholm, Ö = Öland, G = Gotland. Adapted from Aber and Lundqvist (1988, fig. 9).

classic end moraines are now interpreted as ice-shoved features (Moran 1971; Oldale and O'Hara 1984).

In Europe, the outer glaciotectonic zone begins with Ireland, England, and the Netherlands on the west and extends eastward across the southern Baltic and Central European Plain into the Soviet Union. The western and southern Baltic Sea, North Sea, and Norwegian Sea continental shelves are also included. Many case-history examples from Denmark, southernmost Sweden, and the Netherlands are discussed in previous chapters.

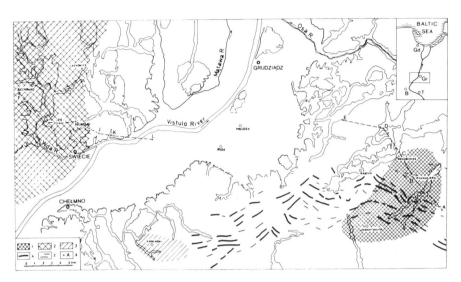

Fig. 9-3. Map showing distribution of buried, ice-shoved hills in the lower Vistula region of Poland. 1, 2 and 3 = glaciotectonic disturbances involving uplifted Tertiary strata; 4 and 5 = major and minor end moraines; 6 = bore-hole sites. Taken from Drozdowski (1981, fig. 1).

Various glaciotectonic phenomena are abundant in northern West and East Germany; deformed chalk in cliffs on the island of Rügen is especially noteworthy. Across Poland, glaciotectonic features are likewise extremely common (Ruszczyńska-Szenajch 1985), including many examples of buried forms (fig. 9-3). Similar glaciotectonic phenomena are presumably present in the Baltic region of the Soviet Union.

The outer glaciotectonic zone of Europe, as in North America, is underlain by relatively soft Mesozoic and Cenozoic sedimentary strata. Cretaceous chalk and Tertiary bedrock of clastic composition are frequently deformed along with preglacial Quaternary sediments. Thick drift, multiple till sequences, and large end moraines are characteristic of this zone. Many of these end moraines have a glaciotectonic genesis (Ber 1987; van Gijssel 1987; Meyer 1987; van der Wateren 1987; Pedersen *et al.* 1988).

The intermediate zone of glaciation presents a startling contrast to the outer zone. In Europe, the intermediate zone includes the Fennoscandian Shield of southern Sweden and Finland and the Caledonian Mountains of Norway and western Sweden. Drift is much thinner and discontinuous, and hard, mostly crystalline bedrock is exposed over large areas. Locally thick drift was left in scattered end moraines and eskers during the last deglaciation, but older drift is only rarely preserved in small, protected sites. Such strata do contain various small (< 10 m vertical) and isolated glaciotectonic structures in Norway (Sønstegaard 1979; Mangerud *et al.* 1981; Haldorsen and Sørenson 1987; Aber and Aarseth 1988) and Sweden (Hillefors 1985; Fernlund 1988).

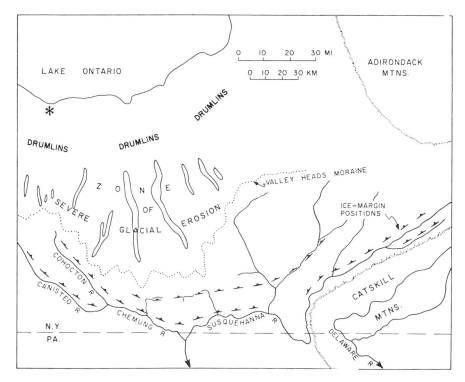

Fig. 9–4. Map of west-central New York showing glacial features. Stratified drift of Valley Heads Moraine contains small ice-pushed structures. Asterisk indicates location of deformed Paleozoic bedrock (Andrews 1980). Map modified from Coates (1974, fig. 8)

The intermediate zone in North America includes much of the Canadian Shield and Appalachian Mountains. Drift cover is thin and patchy, and scoured crystalline rocks outcrop over large areas. Small glaciotectonic structures are occasionally found in locally thick drift, for example in overridden sediment of glacial Lake Merrimack, New Hampshire (Stone and Koteff 1979). Small glaciotectonic features are present on Magdalen Islands in the Gulf of St. Lawrence and on mainland Nova Scotia, where slate, Permian sandstone, and Quaternary sediments are deformed to a depth of 3 m (Dredge and Grant 1987).

In both North America and Europe, the outer and intermediate zones are separated by a transitional belt of varying width, which is underlain by well-consolidated Paleozoic sedimentary bedrock. In North America, this belt begins south of Lake Ontario and stretches westward across the Great Lakes into the Central Lowlands. The belt exhibits gradational features between the outer and intermediate zones. The inner portion shows evidence of strong glacial erosion with relatively thin drift and prominent drumlin fields, but only scattered glaciotectonic features. Toward the outer portion, drift becomes thicker, end moraines are common, and glaciotectonic features are also more abundant.

The outward transition from drumlins and finger-lake troughs formed subglacially to ice-margin features is evident in western New York (fig. 9-4). Ice-contact stratified drift of the Valley Heads Moraine contains small ice-pushed structures related to local oscillations of valley ice tongues. Paleozoic limestone and shale are also locally deformed (Andrews 1980), however such bedrock disturbances are rare.

The outer portion of the transitional belt has more abundant glaciotectonic phenomena, for example in Kansas (chaps. 6 and 7). Similar features in both drift and bedrock are commonly found in adjacent Nebraska (Barbour 1913), Iowa (Lammerson and Dellwig 1957), and Missouri (Howe 1968). Moran (1971) reported several examples of ice-pushed structures from the Great Lakes region. Small hill-hole pairs are present in Wisconsin. Conspicuous glaciotectonic landforms are, otherwise, not common within the transitional belt of Paleozoic bedrock.

The northern limit of the outer glacial zone in Scandinavia is sharply marked in southern Sweden by the Tornquist Line (fig. 9-2), a major structural lineament. Immediately to its north, a mosaic of crystalline and sedimentary bedrock is found in the Fennoscandian Border Zone, which together with Baltic islands of Bornholm, Öland and Gotland (fig. 9-2) make up a transitional belt. Small glaciotectonic structures are present in drift on Öland (Königsson and Linde 1977) and in weakly consolidated Jurassic sandstone on Bornholm.

The inner glaciotectonic zone of Scandinavia extends from south-central Norway, across central and northern Sweden, into northern Finland, and possibly into the Soviet Union (fig. 9-2). Drift is again moderately thick and continuous with drumlins and Rogen moraine common. Exposures of Precambrian and Paleozoic bedrock make up only a small part of the landscape. Interstadial and interglacial sediments are found in many places below till of the last glaciation, and multiple Weichselian till sequences are known (Lundqvist 1967; Hirvas *et al*. 1981; Haldorsen and Sørensen 1987).

Glaciotectonic structures are common within such overridden deposits (Vorren 1979; Lundqvist 1985), and even small ice-shoved hills are found in central Sweden (fig. 9-5). It is clear from the geological context that many of these glaciotectonic features were created in older drift beneath the thick center of the Fennoscandian Ice Sheet (Aber and Lundqvist 1988). The deformation of overridden sediments occurred during all phases of the last glaciation: advancing, maximal, and recessional. Due to migration of the Weichselian ice divide, local directions of ice pushing shifted greatly during the course of glaciation. It is also possible that some glaciotectonic features were formed in ice marginal positions during the final phase of deglaciation.

Two inner zones are present in North America corresponding to two main sectors of the Laurentide Ice Sheet (fig. 9-1). Drift cover is nearly continuous with till plains, Rogen moraine, and drumlin fields in many portions. Precambrian crystalline and sedimentary bedrock is rarely exposed at the surface. These two

Fig. 9–5. Photograph of overturned and thrust, stratified drift that forms the core of a small cupola-hill on Andersön island, near Östersund, central Sweden. See Map 2 for location; photo by J. Lundqvist, 1984.

zones correspond to the major ice divides of the Laurentide Ice Sheet (Prest 1983). Unfortunately, virtually nothing is known about the subsurface structure of the glacial sediments. Based on the central Swedish situation, glaciotectonic phenomena may be relatively common and will probably be discovered with more field work in the inner zones of northern Canada.

A general impression exists that the inner zones in the regions of ice divides were stagnant areas in which the Fennoscandian and Laurentide Ice Sheets had little effect on the landscape, aside from crustal depression and rebound. A few

Table 9–1. Factors considered important for genesis of glaciotectonic phenomena. Compiled from many sources.

		Sub-glacial	Pro-glacial
1.	Lateral pressure gradient	S	P
2.	Elevated ground-water pressure	S	P
3.	Ice advance over permafrost	S	P
4.	Ice advance against topographic obstacle	S	P
5.	Lithologic boundaries in substratum	S	P
6.	Surging of ice lobes	S	P
7.	Subglacial melt-water erosion	S	P
8.	Damming of proglacial lakes		P
9.	Thrusting in front of ice		P
10.	Compressive flow with basal drag	S	
11.	Shearing fault blocks up into ice	S	

geologists have conversely suggested that deep crustal erosion of crystalline basement rocks took place beneath the ice-sheet centers (White 1972). Both points of view appear extreme. It is now apparent that moderate glacial erosion, deposition, and deformation all took place within the inner zones under dynamic conditions of ice movement. The distribution and sizes of glaciotectonic phenomena are related in a general way to availability of deformable strata, namely thick drift or soft sedimentary bedrock. The three-zone model, thus, reflects long-term modification of the substratum during repeated ice-sheet glaciation. Within each zone the local presence or absence of glaciotectonic features may reflect variations in glacier dynamics, permafrost, ground water, lithology of the substratum, *etc.*

Regional Distribution

The regional distribution of glaciotectonic features will be considered for the outer zone, in which such features are abundant and their pattern of distribution may be compared with other glacial phenomena. Various factors have been considered important for controlling the location and distribution of glaciotectonic features (Table 9–1). Basically two regional distribution patterns are recognized for glaciotectonic features:

1. Random, sporatic distribution of megablocks, rafts, diapirs, and other features that have little or no morphologic expressions, along with small cupola-hills. These features were presumably created in subglacial settings far behind ice margins. Their locations are primarily related to conditions of substratum materials.
2. Ice-marginal distribution of morphologically prominent hill-hole pairs, composite ridges, and large cupola-hills. These features were created at or

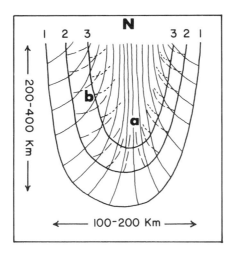

Fig. 9–6. Schematic map showing three phases of an ice lobe. 1 = outer, 2 = intermediate, 3 = inner ice margin positions. Divergent flow lines shown for each phase. Adapted from Chamberlin (1886, fig. 25).

near active ice margins, and their locations are closely related to development of ice lobes or tongues.

The basic geometry of ice lobes was recognized more than a century ago (fig. 9–6). Horberg and Anderson (1956) noted three main factors that controlled the form and extent of Pleistocene ice lobes in the north-central United States: (1) preglacial (bedrock) topography, (2) configuration of the ice sheet, and (3) deflections by adjacent ice lobes. The first of these was undoubtedly most important.

Moran *et al.* (1980) documented the distribution patterns for glaciotectonic phenomena across the Great Plains of western Canada and the north-central United States. Three common associations are apparent; glaciotectonic features are frequently found in conjunction with: (1) bedrock escarpments, (2) subsurface aquifers, and (3) identifiable ice-margin positions. Furthermore, ice margins are often located along bedrock escarpments, because of the topographic control of ice-lobe movement. The large ice-shoved hills along the Missouri Coteau in North Dakota and Saskatchewan are good examples of this combination (Chapter 3).

Ice-shoved hills are arrayed in belts from 2–3 km to 3–5 km wide immediately inside ice margins. These belts define the positions of larger ice lobes. Individual hills within a belt often display their own looped shapes that reflect development of smaller ice tongues related to local valleys or aquifers. Behind these belts, the glaciotectonic hills become generally smaller and smoothed, giving way upglacier to streamlined terrain of low relief (fig. 9–7).

Moran *et al.* (1980) and Bluemle and Clayton (1984) interpreted this basic pattern as simultaneous creation of streamlined and ice-thrust features beneath and behind a stationary ice margin. Thrusting presumably took place in a narrow

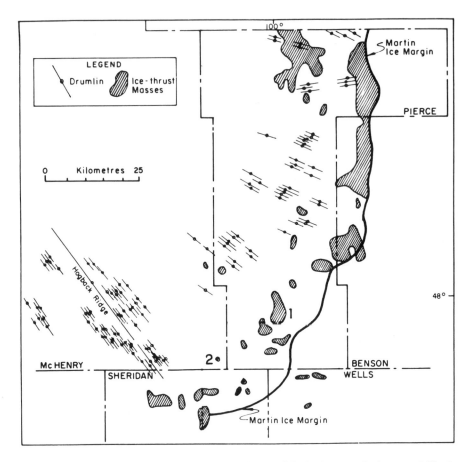

Fig. 9–7. Map of glacial features associated with the Martin ice margin in central North Dakota. 1 = Antelope Hills (chap. 2), 2 = Anamoose (fig. 9–9). Adapted from Bluemle and Clayton (1984, fig. 21); reprinted from *Boreas* by permission of Norwegian University Press, Oslo.

frozen-bed zone, while streamlining occurred upglacier under thawed-bed conditions. Individual thrust blocks were 'plucked' up by the ice. Permafrost was probably involved in many situations, but thrusting of unfrozen material could also occur, particularly above confined aquifers (fig. 9–8). In places where the ice margin advanced, earlier ice-shoved hills were subjected to later streamlining.

Eskers are often associated with ice-shoved hills in central North Dakota, for example at the small hill-hole pair near Anamoose (fig. 9–9). The 30-m-high hill is located immediately southeast of the presumed source depression at Steele Lake. Steele Lake is situated on the margin of a partly buried melt-water trench, which contains a sand-and-gravel aquifer. A small esker begins at the edge of Steele Lake and extends along the northern flank of the ice-shoved hill. The esker was supposedly deposited by ground water flowing from the aquifer, when thrusting

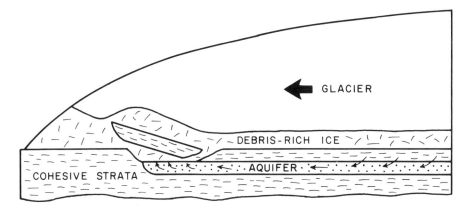

Fig. 9–8. Schematic model for subglacial plucking of large cohesive block above distal end of an aquifer with elevated ground-water pressure. Small arrows show direction of ground-water flow. Adapted from Moran (1971, fig. 5).

opened the downglacier edge of the aquifer. According to Bluemle and Clayton (1984:285):

> The formation of features like those found at Anamoose might be likened to popping the cork from a bottle of champagne; after initial release, the pressure is dissipated. The pressure in the bottle (aquifer) is released as the cork (hill) is removed from the bottle (depression).

The distribution of ice-shoved hills and their relationship to other glacial features in North Dakota is seemingly consistent with subglacial genesis beneath thin, stationary margins of ice lobes. This model is based primarily on the morphologic association of features, because little stratigraphic evidence is available for demonstrating the age or sequence of events.

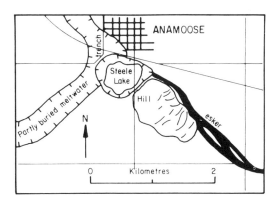

Fig. 9–9. Sketch map of hill-hole pair and esker at Anamoose, North Dakota. For location, see Fig. 9–7; taken from Bluemle and Clayton (1984, fig. 7); reprinted from *Boreas* by permission of Norwegian University Press, Oslo.

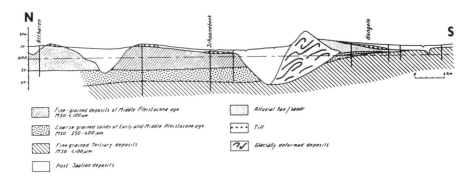

Fig. 9–10. North-south cross section between Schoonebeek and Hengelo in the eastern Netherlands showing position of ice-shoved hill and basin at point where Tertiary strata rise toward the surface and Pleistocene sediments become thinner. Ice movement from north; note large vertical exaggeration. See Map 2 for location; adapted from van den Berg and Beets (1987, fig. 6).

It is equally plausible that ice shoving took place during glacier advance and that the hills were subsequently overridden by ice (Carlson and Freers 1975). The fresh and uneroded appearance of many of the ice-shoved hills on the Great Plains indicates that they were formed late in the deglaciation near the margin of active ice lobes and that subsequent ice movement was short or weak enough to prevent any remolding of the hills.

The lobe-marginal distribution of ice-shoved hills in certain other situations has been interpreted as the result of proglacial thrusting during ice-lobe advance. This pattern of ice pushing is demonstrated best by those cases where ice did not override the hills subsequent to thrusting, for example the Dirt Hills, Saskatchewan (Chapter 3) and Utrecht Ridge, the Netherlands (Chapter 4). In other cases, proglacially thrust hills were later modified by overriding ice, as at the Cactus Hills, Saskatchewan (Chapter 3) and Gay Head, Massachusetts (Chapter 5).

Generally the development of ice lobes, and thus locations of ice-shoved hills, was controlled by preglacial bedrock topography. However, this was not the case in the Netherlands, where the Saalian ice sheet advanced over a relatively flat alluvial plain. Nonetheless, composite-ridges display a well-developed lobate distribution around deep glacial basins that are now partly filled with younger sediment. The creation of these basins was the result of subglacial melt-water erosion combined with glaciotectonic thrusting (van den Berg and Beets 1987).

One more factor related to distribution of glaciotectonic features merits discussion. Ice-shoved hills are frequently found where relatively soft surficial sediments thin or pinch out above hard basement rock. This situation is particularly common in east-central Poland, where soft Tertiary strata become thinner toward the southeast above hard Cretaceous rocks (Ruszczyńska-Szenajch 1978). Ice advance from the northwest created many glaciotectonic disturbances in the zone of Tertiary pinchout. A similar situation is encountered in the easternmost Netherlands, where

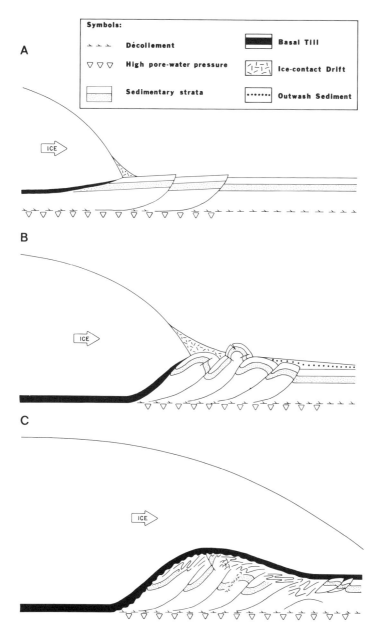

Fig. 9–11. Schematic model for proglacial thrusting and subglacial modification of an ice-shoved hill during glacial advance. Subglacial melt-water features not shown. Modified from Aber (1982, fig. 3).

Tertiary deposits come near the surface. Ice-pushed ridges are found at the position where Pleistocene sediments become thinner (fig. 9–10).

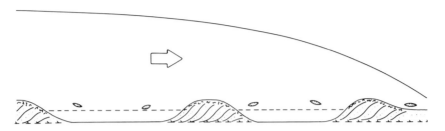

Fig. 9–12. Long profile of idealized glaciotectonic landscape. Not to scale; taken from Aber (1982, fig. 4).

Model for Lobate Pattern of Glaciotectonism

A model for glaciotectonic deformation associated with ice lobes consists of two stages: (1) proglacial thrusting of ice-shoved hills followed by (2) subglacial modification of overridden hills (fig. 9–11). Initial thrusting and shoving up of a composite-ridge take place in front of the ice and above a detachment horizon, or decollement. The decollement may be controlled by several features: lower boundary of permafrost, lithologic boundary, position of confined aquifer, *etc.* High ground-water pressure is presumably developed along the decollement.

Subglacial melt water may either erode tunnel valleys or deposit eskers, while proglacial melt water may erode spillways across the ice-pushed ridges and deposit outwash sediment on the distal side of the hill. Small and temporary lake basins may accumulate sediment between the ice and the hill or between individual ridges. Superglacial debris moving down the ice front may be incorporated into the ice-shoved hill, and basal till may build up on the proximal side. Such features would ideally outline the frontal and lateral margins of the ice lobe at the time of thrusting and may be preserved in those cases where ice did not later overrun the hill.

Continued glacier advance may eventually overrun the ice-shoved hill, at which point a penetrative, metamorphic style of deformation may develop beneath the ice. Erosion of the hill provides reworked sediment for a discordant till cover, and gradually a cupola-hill morphology develops. Further subglacial modification could produce streamlined, crag-and-tail, drumlin forms (Moran *et al.* 1980; van den Berg and Beets 1987). The ice-shoved hill may ultimately be completely destroyed in the subglacial environment.

Under ideal circumstances, a series of ice-shoved hills may be created at different stages during ice-lobe advance and manage to survive throughout the glaciation. This results in a *glaciotectonic landscape* (fig. 9–12), in which narrow belts or loops of ice-shoved hills alternate with wide, low basins. This situation is seen in the Netherlands, where five stages of ice-shoved hills are found (fig. 4–10). According to the interpretation of van den Berg and Beets (1987), these stages occurred during Saalian advance. The older hills (stages e and d) are smaller, and

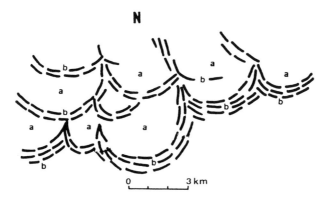

Fig. 9–13. Sketch map showing festoon pattern of ice-shoved ridges (b) and intervening basins (a) for the Suwalki Lakeland, Poland. Adapted from Ber (1987, fig. 6).

the younger (stages c-a) are larger. The older hills were presumably reduced in size by greater subglacial erosion.

Glaciotectonic landscape is also well-developed in the Suwalki Lakeland of northeastern Poland. Ber (1987) described a 'festoon' pattern of composite-ridges, which were formerly interpreted as recessional end moraines, separated by intervening basins (fig. 9–13). He reinterpreted these glaciotectonic features as the results of deformation during ice-lobe advances.

Kineto-stratigraphy

In many regions with abundant glaciotectonic phenomena, it may be difficult or even impossible to conduct glacial stratigraphy in a routine manner. Individual till sheets cannot be traced far. Morphologic features may be relicts of older glaciations. A glacial advance that created much deformation may leave few deposits of its own. In many ways, the problems of stratigraphy in glaciotectonic terrain are analogous to stratigraphy of complexly deformed Precambrian shields. The stratigraphy of shields is often considered as a sequence of deformational events. Each event is marked by a distinctive style, type and orientation of structures and is characterized by certain metamorphic or igneous rocks.

Berthelsen (1973, 1978) developed a similar stratigraphic approach, called *kineto-stratigraphy*, for unraveling complex Quaternary sequences in glaciotectonic terrain. He defined a kineto-stratigraphic drift unit as, 'the sedimentary unit deposited by an ice sheet or stream possessing a characteristic pattern and direction of movement' (Berthelsen 1973:23).

A kineto-stratigraphic drift unit includes all primary sediments (till + stratified drift) that were deposited during all stages of a particular glacier advance and retreat (fig. 9–14). The primary sediments may be deformed by the same ice advance subsequent to deposition. Such deformations are termed *domainal*;

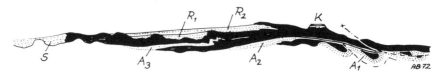

Fig. 9–14. Schematic cross section of deposits classified in a kineto-stratigraphic drift unit. Black = till; stipled = stratified drift. A = advance phase; S = maximum (sandur) phase; R = recessional phases; K = kame. Ice advance from right to left. Taken from Berthelsen (1978, fig. 1).

deformation of subjacent and older strata is *extra-domainal*. The lower boundary of domainal sediments is the limit of penetrative deformation, which also marks the bottom of the kineto-stratigraphic drift unit.

The unifying factor in kineto-stratigraphy is the directional character of the deposits that can be related to ice movements during a particular glaciation. The directional character is revealed by many kinds of features: striations and grooves, till fabrics, indicator erratics, and most importantly glaciotectonic structures. Domainal deformations are most useful in this regard, as there is no doubt about the glacier advance to which they correspond, but extra-domainal structures may also be used for kineto-stratigraphy. The kineto-stratigraphic principle of Berthelsen (1978:29) is that:

> Deposits laid down by successive glaciations can be distinguished by means of the kinetic patterns deduced from the domainal and extra-domainal deformation related to each glaciation, and that the glacial deposits should be correlated according to their directional elements.

In order for kineto-stratigraphy to be applied successfully, three prerequisites must be met. First, consecutive ice advances must come from different directions. Multiple advances from the same direction would all be classified in a single kineto-stratigraphic drift unit. Second, each advance must display a consistent and recognizable pattern of movement. And third, it must be possible to relate various directional features to the proper ice advance. This is a matter for careful field observations and structural analysis.

The second prerequisite – consistent pattern of ice movement – may be the most difficult to deal with, because kineto-stratigraphy is an empirical method. It is tested regionally, but developed by fitting together local results. The local results may at first seem inconsistent or even haphazard when considered in isolation. The direction of ice movement may vary markedly across a region and may even vary significantly at individual sites for different phases of the same glaciation.

Consider an idealized ice lobe advancing from the north (fig. 9–6). A northerly flow direction is developed only along the central axis of the lobe. In the lobe flanks, ice flow diverges toward the southeast and southwest. Thus, flow directions on opposite flanks may differ by as much as 120°. Near the lobe axis (point a, fig. 9–6), flow direction will not change much during the course of glacier advance and

Fig. 9–15. Limits or areas of Weichselian glacial advances in Denmark. 1 = Norwegian advance, 2 = Old Baltic advance, 3 = Main northeast advance, 4 = Young Baltic (East Jylland) advance, and H-R = Hindsholm-Røsnæs (Bælthav) advance. Taken from Berthelsen (1978, fig. 11)

retreat. However, flow directions will shift significantly on lobe flanks as the lobe expands and shrinks (point b). These local patterns of ice movement and glaciotectonic deformation are consistent in light of the lobate model for glaciation.

Kineto-stratigraphy has been most successful for working out glacial advances in the country of its origin – Denmark (Berthelsen 1978; Houmark-Nielsen 1981, 1987). Five phases of Weichselian glaciation are recognized (fig. 9–15). The Norwegian advance came from the north and covered northern Sjælland, Samsø, and Jylland. At about the same time, the Old Baltic ice lobe moved into the southern islands from the southeast.

The Main Weichselian advance next came from the northeast and reached into central Jylland. This advance was complex with at least one or two readvances and a shift to more easterly movement during the course of ice coverage. Following a brief interstade, the Young Baltic ice lobe overspread the islands from the southeast and reached eastern Jylland. The final Bælthav advance took place in the form of ice tongues moving from the south along channels between larger islands. All these advances took place during a relatively short time interval between 20,000 and 13,000 years BP.

The kineto-stratigraphy of Denmark is based on the traditional model for Weichselian glaciation (Holmström 1904; Andersen 1966), in which ice lobes were fed by ice streams following topographic depressions and emanating from the

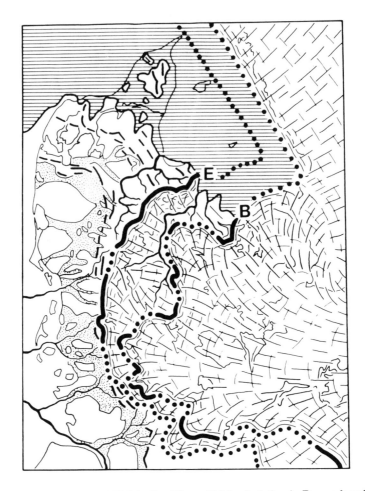

Fig. 9–16. Reconstruction of Weichselian Young Baltic glaciation in Denmark and adjacent areas. E = East Jylland advance; B = Bælthav readvance. Taken from Houmark-Nielsen (1987, fig. 138).

interior of the ice sheet. This model is best illustrated by the Young Baltic phase (fig. 9–16). The ice lobe advanced from east to west along the southwestern Baltic, and divergent ice flow turned toward the northwest and north in southern Denmark. Local variations in ice movement and glaciotectonic deformation could be explained by development of small ice tongues on the margin of the main lobe.

This ice-lobe model for Weichselian glaciation was lately challenged by Lagerlund (1987), who postulated the existence of marginal ice domes. These domes presumably grew during episodes of colder climate on older stagnant ice or inactive surge lobes. The domes developed their own dynamic flow in a radial pattern that was independent of the main Weichselian ice sheet (fig. 9–17). The

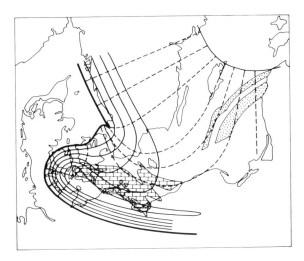

Fig. 9–17. Reconstruction of a marginal ice dome during the Young Baltic glaciation of Denmark and adjacent areas. Taken from Lagerlund *et al.* (1983, fig. 5).

question of ice lobes *versus* marginal ice domes is controversial and is not yet settled (Houmark-Nielsen 1988; Ringberg 1988). Kineto-stratigraphy, through its emphasis on ice movement patterns, will be a valuable approach in this debate.

CHAPTER 10

DYNAMISM OF GLACIOTECTONIC DEFORMATION

Fundamental Cause of Glaciotectonism

It should now be clear that glaciotectonic deformations may take place in a variety of settings – in front of the glacier, beneath the ice margin, or under the center of a thick ice sheet. Deformations may arise during advancing, maximal, or recessional phases of glaciation. All manner of material from well-consolidated bedrock to loose Quaternary sediment in both frozen and thawed conditions may be affected.

Topographic settings vary from rugged mountains to flat plains. Ice-shoved material was usually removed from a basin and piled into a hill of some kind. However, the opposite is also known, where an ice-scooped basin was partly filled with material dislocated from surrounding uplands (Ruszczyńska-Szenajch 1978). In short, glaciotectonic phenomena may be expected wherever sedimentary strata were overridden or pushed by glaciers or ice sheets.

Many factors have been cited as important or necessary conditions for glaciotectonic deformation to take place (Table 9–1). Of these factors, most are related to local topography, substratum material, ice dynamics, or water. These factors may vary considerably over short distances and times. Only the first – lateral pressure gradient – operates everywhere beneath a glacier, regardless of the local nature of substratum material or ice movement. It is the fundamental cause of glaciotectonic deformation (Rotnicki 1976; van der Wateren 1985).

Glaciotectonic deformation takes place when the stress (= pressure) transferred from the glacier exceeds the strength of the stressed material. A glacier imposes two kinds of stress on its bed: (1) vertical stress due to static weight of the ice column (= glaciostatic pressure) and (2) drag or shear stress due to movement of the ice over its bed (= glaciodynamic stress). The glaciostatic pressure is given by ice thickness (H in m) times density (= 0.9 g/cc for ice) divided by 10 (see Table 10–1 for symbols and units):

$$\sigma_{zi} = 0.9H/10 = 0.09H \text{ in kg/cm}^2 \tag{1}$$

This stress equals 90 kg/cm² for each 1000 m of ice thickness. The shear stress created by ice movement is much less, for most situations only 1–2 kg/cm², with maximum values up to 10 kg/cm² where ice is frozen to its bed or flows around bedrock knobs (Weertman 1961; van der Wateren 1985).

The vertical load of ice creates a glaciostatic pressure on the substratum, which is irrespective of ice movement. Assuming constant ice thickness over a level surface, the glaciostatic pressure would be equal and uniform in all directions. However, ice thickness is not constant. Ice thickness varies, particularly near the margin, where thickness increases rapidly from zero at the margin to several 100 m a few km inside the margin. This inequality of ice loading generates a lateral pressure gradient regardless of the direction or rate of ice movement.

The lateral pressure gradient can be calculated easily for situations where ice thickness is known (fig. 10–1). Part of the glaciostatic load on the substratum at any point is transferred to a horizontal stress. The relationship between vertical (σ_{zi}) and horizontal (σ_x) stresses is given by Poisson's ratio (v):

$$\sigma_x = \frac{v}{1-v}(\sigma_{zi}) \qquad (2)$$

For soft sediments and granular materials, Poisson's ratio is close to 0.2, so this value will be used for further calculations (van der Wateren 1985). Thus:

$$\sigma_x = 0.25\sigma_{zi} = 0.0225H \qquad (3)$$

Because the vertical load varies with ice thickness, so the horizontal stress at the glacier sole varies from point to point. The lateral pressure gradient between two points results from the difference (^) in horizontal stresses:

Table 10–1. Symbols and units used in analysis of glaciotectonic deformation.

H	Ice thickness (height in m).
T	Thickness of substratum in m.
P	Pressure or stress: P_t = lithostatic pressure, P_i = intergranular pressure, P_h = hydrostatic pressure.
σ	Normal stress or pressure. Stress unit used in this text is kg/cm². 1 kg/cm² equals approximately 1 bar or 1 atmosphere pressure (= 14.7 psi).
σ_z	Normal stress in vertical direction; stress due to ice loading (= glaciostatic pressure, σ_{zi}), or weight of overburden strata (σ_{zs}).
σ_x	Normal stress in horizontal direction; transverse to ice margin or to ice divide; parallel to slope of ice surface.
σ_n	Stress component oriented perpendicular (normal) to a plane.
σ_{gt}	Glaciotectonic stress operating horziontally at ice/substratum contact. Combination of lateral pressure gradient (Σ ^σ_x) + τ_{ice}.
τ	Shear stress developed parallel to a surface of displacement; measured in kg/cm².
τ_{ice}	Shear stress due to ice movement over substratum (= glaciodynamic stress).
τ_0	Cohesive strength of a material (= cohesion).
θ	Angle of plane (fault) relative to the maximum normal stress.
φ	Angle of internal friction; around 30° for most rocks.
^	Change in value between points of observation, for example: $H_{1/2} = H_1 - H_2$.
v	Poisson's ratio; describes the ratio of horizontal bulging to vertical shortening for vertically stressed rock. Experimental value around 0.2 for most sedimentary rocks.

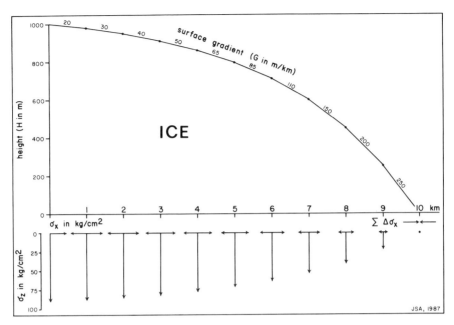

Fig. 10-1. Cross section showing typical ice margin with surface gradient and subglacial pressure conditions for 1-km intervals. The summation of lateral pressure differences at the ice margin equals 22.5 kg/cm², ignoring the drag of ice movement.

$$^\wedge\sigma_{x1/2} = \sigma_{x1} - \sigma_{x2} = 0.0225(H_1 - H_2) \tag{4}$$

For example, consider positions 7 and 8 on the diagram (fig. 10-1) to calculate the horizontal stress difference between two points:

	H	σ_{zi}	σ_x
7	600	54	13.5
8	450	40.5	10.1

$$^\wedge\sigma_{x\,7/8} = \sigma_7 - \sigma_8 = 13.5 - 10.1 = 3.4 \text{ kg/cm}^2, \text{ or} \tag{5}$$

$$^\wedge\sigma_{x\,7/8} = 0.0225\,(H_7 - H_8) = 0.0225\,(150) = 3.4 \text{ kg/cm}^2 \tag{6}$$

The horizontal stress differences are cumulative. In other words, the stress difference over a given interval is passed on and added to the stress difference of the next interval, such that the total horizontal stress difference is:

$$\Sigma\,^\wedge\sigma_x = {^\wedge\sigma_{x\,1/2}} + {^\wedge\sigma_{x\,2/3}} + {^\wedge\sigma_{x\,3/4}} + \ldots \tag{7}$$

In the example (fig. 10-1), the total lateral pressure at the ice margin is 22.5 kg/cm². This stress is independent of ice movement and is controlled entirely by ice load. Lateral pressure should ideally be cumulative over long distance, from the center of an ice sheet to its margin. This cannot actually happen, though, because neither ice nor subglacial strata are ideal materials lacking in cohesion or internal friction. Therefore, the distance over which lateral pressure may build is probably limited to 10 km or less.

The lateral pressure gradient is greatest near the ice margin, where surface gradient is steepest. This explains why most glaciotectonic features are created at or within a few km of ice margins. Maximum cumulative lateral pressure is probably on the order of 25 kg/cm^2, as it would be rare for ice thickness to change by much more than 1000 m over a distance of 10 km.

For ice sheet interiors, thickness changes are naturally much smaller, but are still present transverse to ice divides and do generate small lateral pressure gradients. A thickness change of only 100 m over a distance of 10 km could produce a cumulative lateral pressure of > 2 kg/cm^2. Although smaller in magnitude, interior pressure gradients may also produce glaciotectonic disturbances in appropriate substratum materials.

The total lateral stress in the example (22.5 kg/cm^2) is more than ten times greater than normal subglacial drag and more than twice the maximum subglacial shear stress. It is clear that lateral pressure produced by unequal ice loading is the principal force for causing glaciotectonic deformation in the substratum. To this lateral pressure, the glaciodynamic stress (τ_{ice}) caused by ice movement may be added, assuming ice movement is normally in the same direction as the lateral

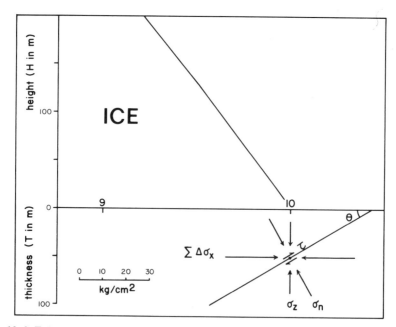

Fig. 10–2. Enlarged view of ice margin in previous figure, showing stresses associated with a plane of potential failure at θ = 30°, 50 m below the ice margin.

pressure gradient. Thus, the total glaciotectonic pressure (σ_{gt}) imposed horizontally on the substratum is given by:

$$\sigma_{gt} = \sum \Delta\sigma_x + \tau_{ice} \tag{8}$$

Given maximum lateral pressure and shear stress values of about 25 kg/cm² and 10 kg/cm² respectively, this means that maximum glaciotectonic stress (σ_{gt}) is around 35 kg/cm². This maximum pressure was realized in only a few restricted situations, for example where a steep ice front advanced upslope over permafrozen ground. Many glaciotectonic features were not created under such unusual conditions; lower horizontal stress of < 20 kg/cm² to only a few kg/cm² was apparently sufficient to produce many glaciotectonic deformations.

Initiation of Thrust Faulting

Thrust faults are the primary structures of ice-shoved hills and megablocks. A subhorizontal thrust, or decollement, separates displaced material above from undisturbed strata below megablocks. Deformed material comprising ice-shoved hills is usually stacked along a series of inclined thrusts. Thus, any attempt to explain glaciotectonic deformation must deal with the initiation of thrusting.

The horizontal pressure caused by glaciostatic plus glaciodynamic stress generates shear stress within the substratum. Shear stress is what initiates thrust faulting. Consider for example a plane of potential failure dipping at 30° and passing 50 m below the ice margin (fig. 10–2). The total lateral pressure gradient is 22.5 kg/cm². We will ignore ice movement, and there is no vertical ice load directly below the margin. The vertical stress is given by thickness (T in m) times density of the overburden strata (2.0 g/cc for water-saturated, unconsolidated sand) divided by 10:

$$\sigma_{zs} = 2.0T/10 = 0.2(50) = 10 \text{ kg/cm}^2 \tag{9}$$

The horizontal and vertical pressures create two stresses on the plane:

1. σ_n – stress oriented normal to the plane.
2. τ – shear stress acting parallel to the plane.

The magnitudes of σ_n and τ can be calculated with the Mohr stress equations (Billings 1972):

$$\sigma_n = \frac{\sigma_x + \sigma_z}{2} - \frac{\sigma_x - \sigma_z}{2}(\cos 2\theta) \tag{10}$$

$$\tau = \frac{\sigma_x - \sigma_z}{2}(\sin 2\theta) \tag{11}$$

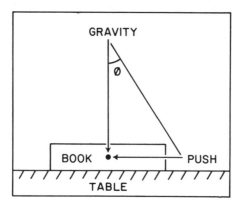

Fig. 10–3. Diagram illustrating angle of friction (ø) and stresses for pushing a book across a table.

Applying these two formulas to the example (fig. 10–2) gives: $\sigma_n = 13.1$ and $\tau = 5.4$ kg/cm². Maximum τ would be achieved on a plane dipping at $\theta = 45°$, so:

$$\tau_{max} = \frac{\sigma_x - \sigma_z}{2} = 6.25 \text{ kg/cm}^2 \tag{12}$$

This explains why most glaciotectonic disturbances occur close to the surface and near to the ice margin. With little overburden or thin ice, σ_z may be small compared to σ_x, and τ_{max} is large. As depth or ice thickness increase, however, so does σ_z, and τ_{max} becomes smaller. When $\sigma_z = \sigma_x$, shear stress is zero, and thrusting cannot develop. Where $\sigma_z > \sigma_x$, the stress system is reoriented (maximum stress is vertical rather than horizontal), and low- to high-angle normal faulting may occur (Croot 1987).

The ideal orientation for thrust faults is at $\theta = 45°$, where maximum shear stress occurs. However, most thrust faults actually develop at about $\theta = 30°$ or less. The reason for this discrepancy is resistance to faulting within the rock mass due to: (1) internal friction and (2) cohesion. These factors can be demonstrated by pushing a book over a horizontal table top (fig. 10–3).

The normal stress acting on the book is simply gravity; shear stress is the horizontal pressure required to move the book. These two stress vectors define an angle (ϕ), called the angle of friction; its magnitude is proportional to frictional resistance. A simple relationship exists between θ and ϕ, such that:

$$\phi = 90 - 2\theta, \quad \text{or} \quad \theta = (90 - \phi)/2 \tag{13}$$

As many rocks possess an internal angle of friction of about 30° (Kulhawy 1975), the θ angle of thrust faulting is also typically around 30°.

Now suppose the book is glued to the table. An extra shear stress will be necessary to break the bond before the book can move. This additional stress is equal to the cohesive strength (τ_0) of the book/table bond. τ_0 values for unconsolidated sediments are essentially zero, whereas most well-consolidated sedimentary rocks have values in the range 250 kg/cm^2 to >300 kg/cm^2 (Kulhawy 1975).

Deformation of better-consolidated rocks actually takes place along weak discontinuities, such as bedding planes, shale seams, claystone or lignite interbeds, and so on. Glaciotectonic pressure is not capable of deforming 'solid' sedimentary rocks. Rather it is unconsolidated sediment or low-cohesion discontinuities within lithified strata that may fail.

According to the Coulomb principle, thrust faulting will occur when the shear stress along a plane of potential failure equals the shearing resistance of the material (Hubbert and Rubey 1959):

$$\tau = \tau_0 + \sigma_n \tan \phi \tag{14}$$

Applying this formula to the example (fig. 10–2): $\phi = 30°$, $\tau_0 =$ zero for unconsolidated sand, and $\sigma_n = 13.1$. τ required for thrusting is 7.6 kg/cm^2, which is more than the actual shear stress acting on the plane (5.4 kg/cm^2). So, thrust faulting could not take place under ordinary conditions.

The mechanism for thrusting large rock masses over long distances was a puzzle to geologists for many years. The necessary physical conditions were elegantly explained by Hubbert and Rubey (1959), and Mathews and Mackay (1960) quickly applied this explanation to ice-shoved features. The total lithostatic pressure (P_t) exerted at any level within the substratum below a glacier results from weight of the overburden strata plus the ice cover ($\sigma_{zs} + \sigma_{zi}$).

This lithostatic pressure is comprised of two components: (1) intergranular pressure (P_i) and (2) hydrostatic pressure (P_h). Intergranular pressure is a mechanical stress transmitted from solid grain to grain within the rock mass. Hydrostatic pressure is a fluid stress transmitted through the water column within pores and fractures in the rock mass. Below a glacier, the substratum is assumed to be completely saturated with ground water, so substratum pressure can be represented as:

$$P_t = P_i + P_h \tag{15}$$

Hubbert and Rubey (1959) demonstrated that the critical shear stress for movement along a thrust fault depends only on intergranular pressure, such that:

$$\tau = P_i \tan \phi, \quad \text{or} \quad \tau = (P_t - P_h) \tan \phi \tag{16}$$

Lithostatic pressure is normally supported largely by intergranular pressure, and hydrostatic pressure is less than lithostatic pressure. However, in some situations hydrostatic pressure may approach or equal lithostatic pressure, in which case the overburden material essentially floats on a high-pressure fluid cushion that has no cohesive strength ($\tau_0 = 0$). As hydrostatic pressure reaches lithostatic pressure, the shear stress required for thrust faulting approaches zero.

In a glacial context, elevated hydrostatic pressure can develop in two manners:

(1) compaction of incompetent and impermeable strata, such as claystone or lignite, and (2) transmission of water into a confined aquifer under a pressure head. The former occurs wherever glacier advance loads a sequence of interlayered sediments, and the incompetent beds are compacted faster than water can escape. The latter is often observed around modern glaciers, where fountains, jets, and sudden floods of melt water emerge from below the ice margin (Gustavson and Boothroyd 1987).

Suppose that a confined aquifer exists below the ice margin in the example (fig. 10–2), and this aquifer is fed from an upglacier water table within the ice 100 m above the substratum. This extra water head is equal to height of the water above the substratum (in m) times the density of water divided by 10:

$$P_h = 100(1)/10 = 10 \text{ kg/cm}^2 \tag{17}$$

For a given plane, P_t is equal to σ_n, and we will assume τ_0 is zero at the time of thrusting. Thus:

$$\tau = (\sigma_n - P_h) \tan \phi \tag{18}$$

For the example (fig. 10–2), this gives:

$$\tau = (13.1 - 10) \tan 30 = 1.8 \text{ kg/cm}^2 \tag{19}$$

Thrust faulting would take place readily under these conditions of shear stress, hydrostatic pressure, and material cohesion. No ice movement is required; the lateral pressure gradient combined with elevated ground-water pressure is sufficient to initiate thrust faulting. Any ice flow toward the margin simply augments the available shear stress.

Thrust faults ideally develop at about $\theta = 30°$ within homogeneous material. However, the position of thrusting is often controlled by a pre-existing weakness within the substratum: bedding plane, incompetent bed, lithologic boundary, unconformity, permafrost boundary, *etc.* Such discontinuities are approximately horizontal in many sedimentary sequences, so subhorizontal thrust faults are quite common.

Continuation of Thrust Faulting

The continued movement along a thrust fault requires maintenance of the critical shear stress and may involve overcoming increasing resistance to fault movement. Consider a flat megablock moving horizontally over level terrain beneath a glacier (fig. 10–4). As long as the pressure conditions that caused initial thrusting continue to exist, the megablock may be displaced horizontally.

Long-distance transportation over glaciated plains is possible (Chapter 6). Many megablocks in western Canada were moved over and deposited on till of the same glaciation. Movement will cease when either hydrostatic pressure or glaciotectonic

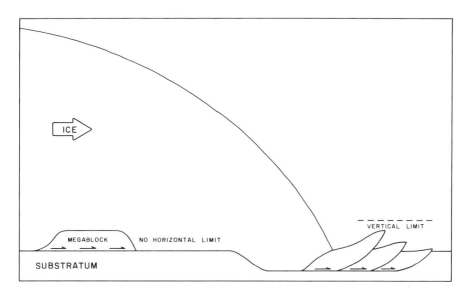

Fig. 10–4. Schematic diagram illustrating continued displacement of thrust blocks in two settings: horizontal subglacial megablock and inclined ice-marginal thrusts. Not to scale.

stress decreases, for whatever reason, or a topographic obstacle is encountered. Most megablocks were probably moved as individuals; that is displacement of one megablock was independent of other megablocks in time and space.

Continued movement on dipping thrust faults poses a different problem, for here the fault block is pushed up an inclined ramp (fig. 10–4). Part of the thrust block – its toe – must be lifted above the original surface; this is what creates an ice-shoved hill. The weight of the thrust toe represents increased resistance to fault movement, and the toe weight grows as thrusting continues. Thus, shear stress must be large enough to overcome initial shear resistance on the fault as well as to lift the thrust toe (van der Wateren 1985).

The effect of toe weight explains several common observations about ice-shoved hills. First, the amount of vertical uplift is limited, for large hills to between 100 and 150 m, with maximum vertical uplift of 200 m (Kupsch 1962). Given a thrust-plane dip of 30°, this means only around 200 to 350 m of horizontal displacement along the ramp itself. This may at first seem like too small a figure for horizontal movement, but the ramp may itself also be displaced.

The second effect of toe weight is to increase the load on the substratum below the front end of the thrust block. This will initiate a second, more distal thrust fault (fig. 10–4). When thrusting of the first block ceases, it will be carried forward on the back of the second thrust block, and so on, in a telescoping manner. In this way, composite-ridges are constructed, in which proximal thrust blocks may be moved some km, whereas the most distal will move only a short distance.

During thrusting of composite-ridges, individual thrust blocks undergo various internal deformations and may be pushed into steeper positions as they pile up in front of an advancing ice margin. Eventually the combined toe weights and internal resistance to thrusting of the composite-ridge will exceed the glaciotectonic stress and thrusting will cease.

Another effect of toe weight is to cause collapse of the uplifted part of the thrust block. Once raised above the original surface, lateral support for the toe is lost, and the maximum stress (gravity) is vertical. Normal faults, slumps, fissures, and related structures may form, as a result of relaxation of glaciotectonic stress. Such secondary collapse structures are most common in the upper portions of ice-shoved hills and may be the only structures visible in shallow exposures. Some care for interpretation is advisable, as these structures may not represent the primary genesis of an ice-shoved hill.

The final effect of toe weight is to place severe limitation on subglacial ramp thrusting. In addition to lifting the toe weight, glaciotectonic stress must also displace the column of ice above the toe. It is simple enough to say that subglacial thrust blocks may be 'incorporated' into the glacier (Moran 1971). However, it is mechanically very difficult to develop sufficient stress to lift large thrust blocks beneath a glacier.

Glaciostatic pressure, which is by far the greatest stress acting at the glacier sole, prevents uplift of more than a few 10s of m. Subhorizontal thrusting or development of low-angle normal faults is to be expected, rather than ramp thrusting, in subglacial settings (Croot 1987). So, whereas megablocks and small ice-shoved hills may be created subglacially, large hills can originate only at ice margins.

Scale Models of Glaciotectonism

Large ice-shoved hills are not actively forming anywhere today, and so we have no modern example for comparison with Pleistocene features. This is part of the reason for the variety of opinions on genesis of large Pleistocene ice-shoved hills. Fortunately, some scale models, both natural and artificial, may be relevant to ice-shoved hills in general. Gry (1940) made an early attempt to model ice-push deformation by applying horizontal pressure on snow layers. He generated imbricated thrusts, which rotated into steeper positions as pushing continued.

A major difficulty for artificial models is correctly scaling down all physical factors, from sediment characteristics to rate of deformation, in order to produce a realistic simulation. Mulugeta and Koyi (1987) have beautifully modeled the 'piggyback' style of thrusting. They subjected thin (0.1 to 0.2 mm) layers of fine (0.08 to 0.18 mm) sand to 40% bulk shortening in a conventional squeeze box. The deformed sand layers were then exposed by using a small vacuum nozzle to 'erode' into the thrust mass. Three domains of deformation result under these conditions (fig. 10–5).

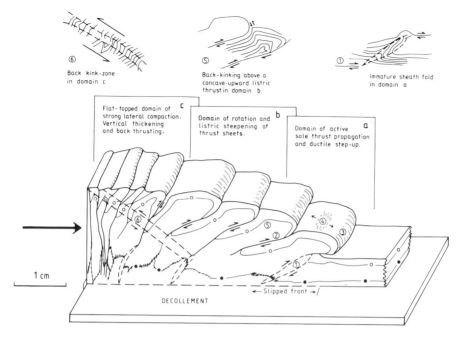

Fig. 10–5. Deformation of stratified fine sand subjected to 40% lateral shortening in a squeeze box: 0 = individual layers, 1 = initiation of thrust, 2 = thrust fault, 3 = slump zone, 4 = extension fractures, 5 = back kink fold, 6 = back thrust zone. Taken from Mulugeta and Koyi (1987, fig. 3).

The distal domain (a) consists of low-angle thrusts, between which thrust blocks develop overturned drag folds, extension fractures, and slumps. Thrust blocks are rotated into steeper positions in the intermediate domain (b), and back kinking develops in the upper folded portions of thrust blocks. The most severe deformation occurs in the proximal domain (c), where thrust blocks are rotated into near-vertical positions and are laterally compacted. Back thrusts or kink zones, that dip outward, result in underthrusting and further thickening in the proximal domain.

This model was developed to analyze and demonstrate the kinematics of piggyback thrusting, not as an analog for a specific type of thrust terrain, so it represents only an approximation of glaciotectonic structures. Most ice-shoved hills exhibit domain a and b features. Domain c features imply much greater uplift and lateral compression than is usually developed, although such features are present in the core portions of some composite-ridges.

The model deals strictly with deformation of pre-existing strata. In glaciotectonic settings, the ice and melt water also deposit new sediments, which may become intermingled with older material as deformation proceeds. Furthermore, continued glacier advance may overrun the ice-pushed hill, at which point a new stress regime combined with glacial erosion or deposition would be imposed.

The only true scale models of ice-shoved hills are natural ones that we can observe in connection with modern glaciers. Humlum (1983) described a small

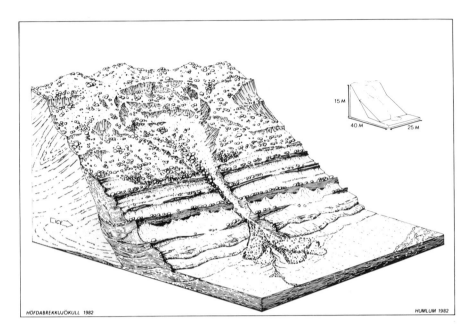

Fig. 10–6. Block diagram showing surface expression and internal structure of push-moraine forming in front of Höfdabrekkujökull, Iceland. Taken from Humlum (1983, fig. 5).

push-moraine forming in front of Höfdabrekkujökull, an advancing outlet glacier of Mýrdalsjökull, Iceland (Map 2). The moraine is composed partly of unfrozen floes thrust up from proglacial outwash and partly of superglacial debris and mud-flow deposits (fig. 10–6). The moraine is 5–10 m high and has a terraced front where individual floes protrude.

Initial thrusting took place in front of the advancing ice margin. The average angle of thrusting (θ) is 26°, giving an angle of friction (ϕ) of 38° for the sand and

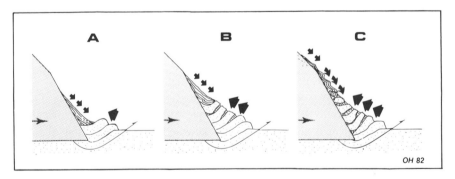

Fig. 10–7. Model for sequential creation of push-moraine at Höfdabrekkujökull, Iceland. Thrust floes are pushed up the ice front, while superglacial debris moves downward. Taken from Humlum (1983, fig. 11).

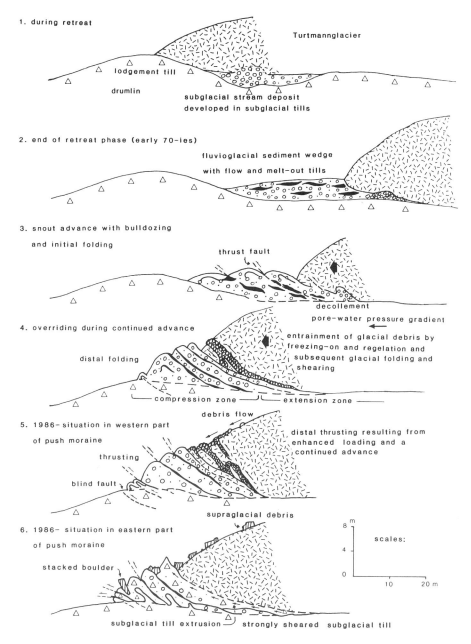

Fig. 10–8. Idealized sequence for creation of the push-moraine in front of Turtmannglacier, Switzerland. Taken from Eybergen (1987, fig. 10).

gravel sediment. This larger-than-usual angle of friction is because of the volcanic origin of the sediment; the grains are angular in shape and thus abrasive.

Once formed, individual floes were pushed up the ice front, as new floes appeared beneath, in a sequential manner (fig. 10–7). Meanwhile, superglacial debris moved down the ice front, which resulted in composite interlayering of material in the push-moraine. This situation demonstrates proglacial thrusting of unfrozen and unconsolidated sediment with a rather high angle of friction. Given its small size, this push-moraine's chances for long-term preservation are probably slight, however.

Eybergen (1987) has investigated a push-moraine complex created in front of the Turtmannglacier in Wallis, southern Switzerland (Map 2). The Turtmannglacier is a typical Alpine glacier, which readvanced about 100 m during the 1971–86 period. The push-moraine is made up of small composite-ridges, 5–11 m high, developed in lobate plan more-or-less parallel to the glacier snout. It is 100 m long east-west and nearly 40 m wide. The push-moraine is composed of till plus coarse, sandy gravel and rests on a drumlin of older lodgment till.

The eastern portion of the push-moraine contains mainly till derived from deformation of older lodgment till and squeezing of material from beneath the ice margin. Gravel dominates in the western portion; the gravel is derived from thrusting of at least partly frozen older outwash deposits and melting out of englacial and subglacial debris. Volume of the push moraine is estimated at 12,000 m^3, of which < 1000 m^3 is material supplied from the glacier during the period of readvance (Eybergen 1987). Hence, the moraine is primarily a glaciotectonic feature consisting largely of deformed, older glacial sediments.

During readvance of the Turtmannglacier, proglacial thrusting took place in both till and outwash gravel (fig. 10-8). Thrusting was facilitated by development of increased pore-water pressure within the overridden sediment and in confined beds of the sediment wedge in front of the glacier. Piggyback-style thrusting migrated outward, with youngest deformation in the most distal portion. The structures correspond to domain a and b deformation. Continued strong advance could destroy the push moraine. If the Turtmannglacier should recede, however, the moraine would be protected from melt-water erosion because of its position on a drumlin.

The various natural and artificial scale models demonstrate the kinds of structures and morphology that may be expected in larger ice-shoved hills, where both drift and bedrock were deformed. Most ice-shoved hills were apparently created close to or in front of ice margins, and therefore other ice-marginal processes must be taken into account when interpreting the origin of such hills. Megablocks and some small ice-shoved hills were formed subglacially. In many cases, features that originated in proglacial settings were later modified by overriding ice. The conditions of subsurface pressure, particularly hydrostatic pressure, were of paramount importance for genesis of glaciotectonic phenomena.

CHAPTER 11

GLACIOTECTONIC ANALOGS

Introduction

Many structures within the earth's crust are created in a manner analogous to glaciotectonic deformation. Wherever an advancing mass imposes an increasing load on weak substratum material, the conditions for deformation may develop both in front of the mass as well as beneath it. Glacier ice is only one kind of mass whose distribution and resulting crustal load may increase through time. Local increase of crustal loading may come about in many other non-glacial situations: growth of an alluvial fan or a delta, buildup of volcanic deposits, landslide or mass movement of material, and convergence of lithospheric plates.

The scale of these analogous structures ranges from superficial cm-size deformations in soft sediments to major mountain ranges. In a general way, the size of structures and depth of disturbance are related to the size of the advancing load and the time interval during which that load was effective in reducing shear strength of the deformed material. Displacement above a decollement with elevated hydrostatic pressure is an essential ingredient in all situations from ice pushing of small composite-ridges to thrusting of major mountain ranges. Glaciotectonic deformation is not unique from this point of view; it is similar to many other crustal disturbances, the only significant differences being temporal and spatial scales.

Discussion of analogous non-glacial deformation will provide information and ideas relevant to further understanding the genesis of ice-pushed structures. This is a two-way exchange, for knowledge of glaciotectonic features could likewise contribute to the interpretation of non-glacial structural deformation. Three glaciotectonic analogs of increasing size are presented in this chapter. First are mudlumps of the Mississippi Delta, which are similar in size to large composite-ridges. Glaciotectonic structures are next compared with thin-skinned thrusting of mountains, and finally similarities between convergent plate-boundary and ice-push deformation are discussed.

Mississippi Delta Mudlumps, Mississippi, United States

The Mississippi Delta is a classic, large, bird-foot style delta (fig. 11–1). One of its lesser known features are mudlumps – small islands or shoals formed over uplifted clay structures near bar-finger sands at the mouths of major distributary channels (Coleman 1988). Mudlumps are active features with life spans lasting a few decades before they become buried by prograding delta sand. The zone of mud-

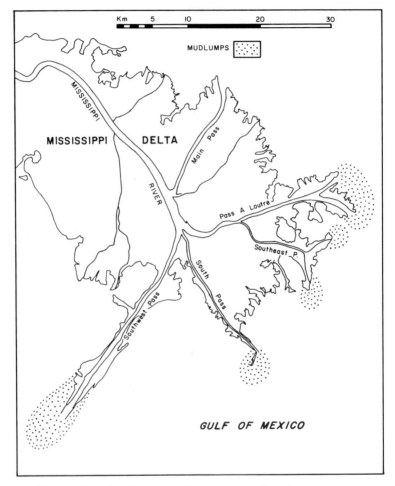

Fig. 11–1. Map of Mississippi Delta showing larger distributaries and zones of mudlump development. Based on Morgan *et al.* (1968, fig. 1).

lump uplift has migrated outward > 2 km during the past century at South Pass distributary. This corresponds to southward extension of the distributary-mouth bar. Morgan *et al.* (1968) concluded that exposed mudlumps are spines on a series of linear diapiric folds that developed peripherally to the deforming load of the bar.

Mudlumps range in size from pinnacles to small islands with maximum areas of about 8 hectares. They are mostly oval in form, with length usually 3–4 times the width. The surficial portion of a typical mudlump is formed by an anticline trending parallel to the mudlump's long axis (fig. 11–2). A narrow, shallow graben marked by many small normal faults runs along the crest of the anticline. Strata within the graben are highly irregular and confused. Extrusion of mud volcanoes and venting of methane-rich gas during uplift indicate that excess hydrostatic pressure must be developed within the clay cores of mudlumps.

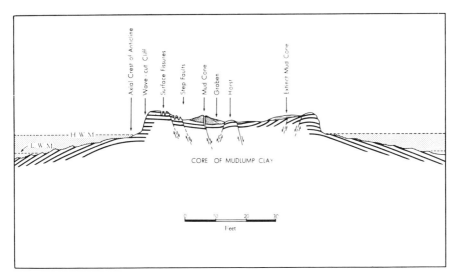

Fig. 11–2. Schematic, transverse cross section showing surficial features of a typical mudlump. H.W.M. = high water mark; L.W.M. = low water mark; scale in feet (30 feet = 9.2 m). Taken from Morgan (1961, fig. 6).

Mudlumps have formed only at the mouths of major distributaries; they have not developed in connection with lesser or shallow distributaries. Mudlumps undergo three developmental stages: (1) initial uplift as a submarine shoal, (2) growth into an island, and (3) erosion and truncation by waves. Many mudlumps experience episodic uplift, which invariably coincides with river flooding, when rapid sedimentation occurs at the distributary mouth. Some mudlumps have risen 3–5 m during a single flood cycle (Coleman 1988). It is therefore concluded that mudlump

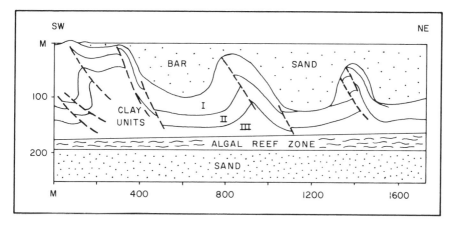

Fig. 11–3. Subsurface section located southwest of South Pass distributary. Thrusts and overturned folds in prodelta clay strata are present beneath mudlumps. Vertical exaggeration = 2.5; adapted from Morgan *et al.* (1968, fig. 9).

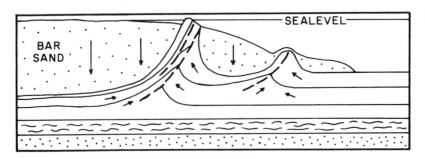

Fig. 11-4. Schematic model for structural development of mudlumps during advance of delta sand from left to right. Short arrows show direction of clay flowage; long arrows indicate differential loading by sand. Symbols same as Fig. 11–3; adapted from Morgan *et al.* (1968, fig. 23F).

development is intimately connected with the sediment load created by seaward growth of distributaries (Morgan *et al.* 1968).

Test drilling near South Pass distributary has revealed the subsurface stratigraphy and structure of several mudlumps (fig. 11–3). Beneath bar-finger sand, which may reach up to 120 m in thickness, a sequence of prodelta clay units rests on an algal reef zone 180 m deep. The algal reef zone is dated at 26,500 years BP, and a shell layer between clay units I and II is dated at 15,500 years BP. These dates indicate a late Wisconsin age for the prodelta sediments that were deposited during the late-and post-glacial rise in sea level.

The clay units have been deformed into a series of asymmetrically thrust anticlines or diapirs with the algal reef zone acting as a structural basement. Clay strata exposed in mudlump islands contain foraminifera derived from at least 120 m or more in depth (Andersen 1961). The mudlumps of South Pass distributary have developed during the past 100 years, and thick sand has built up in the subsiding synclines between mudlumps during the same time. Assuming distributary sand accumulates near sea level, then synclines have subsided at average rates > 1 m/year (Morgan *et al.* 1968). This subsidence compensates for uplift in mudlumps. The asymmetry of folds, faults, and diapirs reflects differential loading by the accumulating sand mass.

Mudlump uplift proceeds sequentially in front of advancing distributary-bar sand in a manner akin to piggyback thrusting (fig. 11–4). However, in this case no horizontal pushing or drag is exerted by the sand on underlying clay strata. The driving force for deformation is the lateral pressure gradient produced by unequal loading of relatively heavy sand over relatively light and plastic clay. Mudlump uplift continues until the original source of clay is depleted or until the mudlump is buried by sand. The distribution of active mudlumps follows lobate patterns around the ends of prograding distributaries.

The size, style, and pattern of Mississippi mudlump deformation is quite comparable to large composite-ridges. Advancing distributary-bar sand serves the

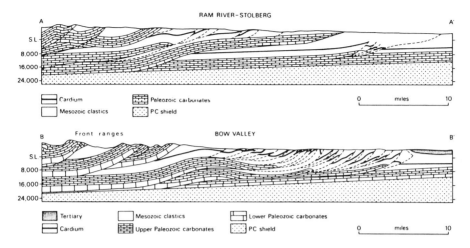

Fig. 11–5. Cross sections through the eastern part of the Canadian Rocky Mountains. Adapted from Bally *et al.* (1966).

same role as advancing ice. In both cases, static loading generates lateral pressure gradients that are responsible for deformation around the frontal and lateral margins of the ice or sand. The advancing load may eventually overspread the earlier-deformed structures. Mississippi mudlumps demonstrate that features similar to large composite-ridges can be created entirely by differential loading in a totally thawed environment (Aber 1988f).

Thin-Skinned Thrusting

There are many examples of tectonic deformation involving displacement of surface sedimentary strata over basement rocks by orogenic processes. The most intensively studied mountains are probably the Canadian Rockies (Bally *et al.* 1966; Dahlstrom 1970, 1977; Perry *et al.* 1984), the Appalachians (Rich 1934; Harris and Milici 1977; Perry *et al.* 1984), and the Alps (Graham 1978; Lemoine 1973; Siddans 1979, 1984).

At a descriptive level, many close similarities exist between glaciotectonic landforms, particularly composite-ridges, and orogenic-scale thrust masses. These similarities extend beyond the morphological comparisons made by Gripp (1929), to similarities of internal structures, geometry of the systems, and models proposed to explain their genesis. Lowell (1985) provides an excellent summary of geologic structures, including sectional data from a wide variety of regional settings (fig. 11–5).

Many composite-ridge systems are composed of imbricated sheets of sediment displaced by glacial activity from flat-lying strata just beyond the ice margin. Such imbrication is also one of the most commonly observed features of many Alpine

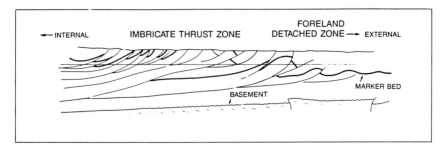

Fig. 11–6. Idealized thrust-fold belt with imbricated thrust zone and foreland detached zone (after Lowell 1985).

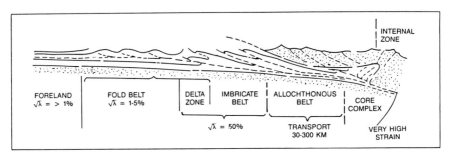

Fig. 11–7. Idealized thrust-fold belt with percentage strain/shortening in each zone. From Roeder (1983).

foreland zones and gives rise to a starting point for the thin-skinned thrust analog (figs. 11–6, 11–7). Other common elements include:

1. Major thrust surfaces, which are bed-parallel in less competent strata forming 'flats,' and which cut upsection in competent units forming 'ramps' (Harris and Milici 1977; fig. 11–8).
2. A common plane of decollement, usually a weak zone or horizon, from which all thrusts propagate upward (Boyer and Elliott 1982).
3. Individual thrust sheets that are laterally discontinuous (or of limited axial length) and form ridges that rise and die out again.

As in glaciotectonic settings, it is common to find orogenic thrusting and folding associated together in the same system, with complex spatial and temporal relationships between each style of deformation (Dahlstrom 1977; Lowell 1985). Siddans (1979) remarked on the limited axial length of arcuate folds and thrusts associated with the European Alpine foreland in France. A great deal of further research needs to be carried out on the basic similarities in the two systems. Nevertheless, the overlap between styles and geometries of movement enable us to look beyond the level of descriptive comparison of individual elements to whole thrust systems and to a variety of models for their development (Table 11–1).

Table 11–1. Structural styles, dominant stresses, and transport modes for thin-skinned deformation in glacial settings. Modified from Harding and Lowell (1979) and Lowell (1985).

Structural Style	Dominant Stress	Transport Mode	Glacial Setting
Decollement thrust-fault assemblages.	Compression	Subhorizontal to high-angle convergent dip slip of sediment in slabs, sheets and lobes.	Advancing glacier in foreland, flanks or valley side.
Detached normal-fault assemblages.	Extension	Subhorizontal to high-angle divergent dip slip of sediment in sheets, wedges and lobes.	Passive collapse or overriding by glacier.
Diapiric structures.	Vertical loading	Vertical and horizontal flow of fluid sediments, arching or piercing of cover sediments.	Unstable density stratification below glacier.

Geologists currently employ three main mechanistic models to explain the development of thrust systems: (1) gravity gliding/sliding, (2) gravity spreading, and (3) pushing from the rear (compression). The applicability of such models has been vigorously debated by Elliott (1976), Chapple (1978), and Siddans (1984) among others.

Pedersen (1987) provided a useful introduction to the debate from an outside viewpoint. Lowell (1985) added a further dimension to the debate concerning the mechanics of geologic thrust-fold-belt assemblages. He argued that none of these models apply and that underthrusting is the only viable model. Underthrusting cannot, however, be used to account for most glaciotectonic structures, and so we must look to the aforementioned models for analogs.

Gravity gliding/sliding is a mechanism in which deformation is caused by movement of the sediment package downslope under its own weight. In a glaciotectonic setting, such a model implies that deformation of sedimentary strata in front of an advancing glacier occurred because the proximal zone was uplifted, for example as a forebulge, providing a surface slope away from the glacier (fig. 11-9). As far as we are aware, very few structures of this origin are observed in glaciotectonic settings.

Gravity spreading is also induced by a change in gravitational forces acting on a once stable sediment package. In this case, a normal load applied by an increased weight (ice, sediment or rock) is translated into lateral stresses which result in pushing of the adjacent sediment away from the load and upward (Elliott 1976).

176   CHAPTER 11

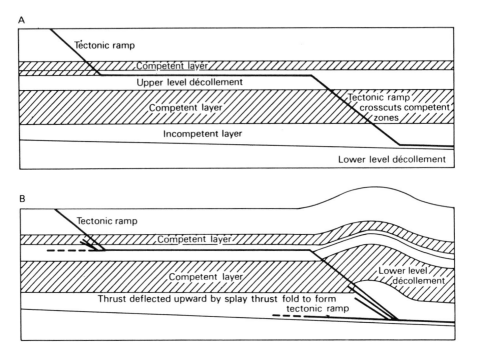

Fig. 11–8. Common style of development of ramps and flats, where thrusts are bed-parallel in weak or incompetent zones and cut upsection through competent layers. A = pre-deformation; B = post-deformation.

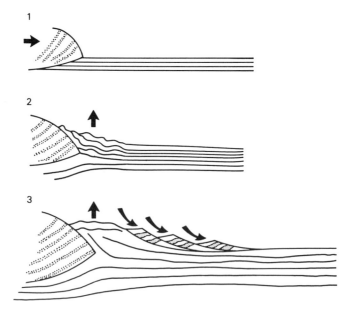

Fig. 11–9. Gravity-sliding model in a glaciotectonic setting.

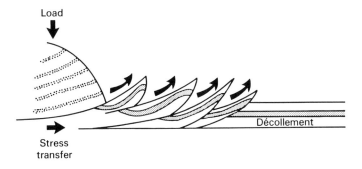

Fig. 11–10. Gravity-spreading model in a glaciotectonic setting.

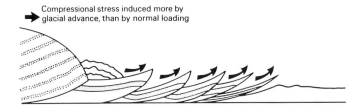

Fig. 11–11. Pushing-from-the-rear model in a glaciotectonic setting.

Such increases in weight may be caused by glacier loading, sediment buildup, orogenic loading, or even by dumping man-made waste (Ruszczyńska-Szenajch 1986).

The style of deformation in sediments adjacent to the load point is listric (curved, concave-upward) thrusting that dies out with distance from the load. In a glaciotectonic setting the implication is that the major stress is due to ice loading (normal stress) not drag created by ice advance (Rotnicki 1976). The main principles are illustrated in Fig. 11–10 (see also fig. 10–1).

The final model, pushing from the rear, was proposed by Chapple (1978) as an explanatory model for structures identical in style to those that Elliott (1976) accounted for by gravity spreading. Pushing-from-the-rear models, which at first sight appear to be most logical, have met with considerable criticism from structural geologists. Many have consistently argued that thin, broad sheets of sediment cannot transmit sufficient stress to remain undeformed while undergoing thrusting.

Chapple's (1978) mechanistic treatment appears to contradict this, and he argues that 'compressive stress (by a pushing movement) is a fundamental feature of thrust belts.' (1978:1196) His model includes a weak basal layer and a wedge-shaped sedimentary package. The implication for glaciotectonic deformation is that the forward motion of the glacier is critical, not the static ice load (fig. 11–11).

Siddans (1984) considered each of these models in different configurations of surface slope and sediment package in an Alpine orogenic setting (fig. 11–12). He

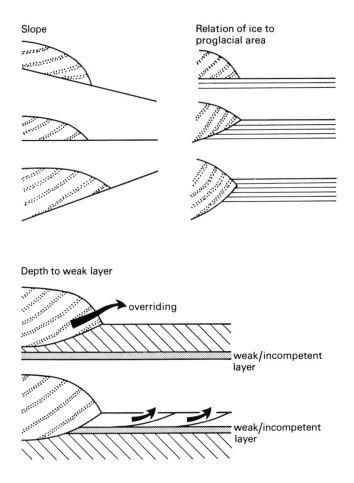

Fig. 11–12. Diagrammatic illustration of the adaptation of parameters considered to be important for modeling orogenic thrust-fold belts in relation to glaciotectonic settings.

argued that each model is mechanistically possible and could produce tectonic deformation, providing the underlying assumptions are fulfilled. In a glacial setting, the most critical factors appear to be (fig. 11–12):

1. Slope of the proglacial area
2. Depth to a suitable weak layer (decollement)
3. Relationship of glacier snout to proglacial sediment
4. Thickening of the proximal end of the sediment package

Croot (1987) argued that the pushing-from-the-rear model explains the piggyback development of composite-ridges in Iceland, whereas van der Wateren (1985) and Pedersen (1987) considered the gravity-spreading model most appropriate for the same style of glaciotectonic deformation. In all likelihood, pushing from the rear and gravity spreading operate together to a greater or lesser degree, depending on

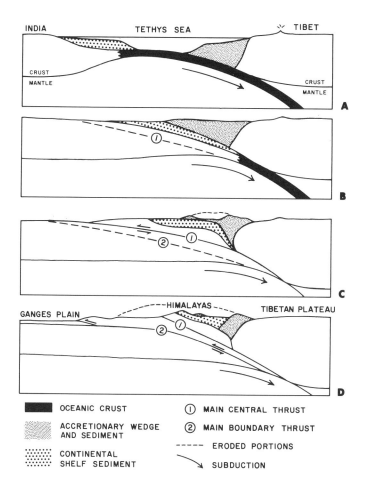

Fig. 11–13. Schematic cross sections showing developmental stages of the Himalayas and Tibet Plateau: A – 60 million years BP, B – 50 million years BP, C – 35 million years BP, D – 20 to 10 million years BP. The Himalayas have been thrust southward as India underthrust Tibet. Adapted from Molnar (1986, p. 74); vertical dimension approximately 75 km.

the factors listed above. So, many composite-ridges may be considered as results of both mechanisms.

Modeling of glaciotectonic deformation is still in an early stage of development. It seems probable that glaciotectonics may provide more rapid solutions to modeling of orogenic thrust systems than *vice versa*. Each area of geologic study can benefit from developments in the other.

Convergent Plate Boundary

The earth's largest topographic features–mid-ocean ridges, deep-sea trenches, and mountains–are the results of plate tectonics. The rigid outer shell of the earth, called the *lithosphere*, moves slowly as several large and many small plates, whose sizes, shapes and positions are constantly changing. The lithosphere is composed of the crust plus the uppermost mantle with an average thickness of 100 km, thicker under continents and thinner under oceans.

Plates interact along their mutual boundaries, which may be either divergent, convergent or transform. Mountain systems are created where convergent boundaries develop within or next to continental lithosphere. Convergent boundaries involving continents take two general forms (Molnar 1986):

1. Subduction zone – where thin oceanic lithosphere descends below a continental plate and a volcanic mountain chain develops on the overriding continent. Such mountains are supported by deep but relatively warm and weak crustal roots. The Andes Mountains are a good example.
2. Collision zone – where thicker oceanic or continental lithosphere moves into a subduction zone and thrusts under the overriding continental plate. Such mountains are held up by thinner, but colder and stronger crust. The Alps of central Europe are a good example.

The collision-zone type of convergent boundary is comparable in a general way to ice-pushing of composite-ridges. Ignoring the great differences in time span of development, lithology of affected rocks, and size of structures, an overriding continent behaves like an advancing ice sheet: where an obstacle is encountered, thrusting of that obstacle may occur in front of and below the leading edge of the continent or ice sheet. Both types of convergent plate boundaries as well as the glaciotectonic analogy are demonstrated by development of the Himalaya Mountains and Tibet Plateau.

At the beginning of the Cenozoic (fig. 11–13A), India was moving rapidly northward approaching Asia. The two continents were separated by an ocean basin – the Tethys Sea, which was closing due to subduction beneath Asia. Volcanic mountains grew above the subduction zone and the crust was greatly thickened by intrusions to form most of what is now the Tibet Plateau. Along the southern edge of Tibet, land-derived sediments along with oceanic material scraped off the subducting plate built up an accretionary-wedge complex. Meanwhile, continental-shelf sediments continued to accumulate along the northern edge of India.

Approximately 50 million years ago (fig. 11–13B), India began to move into the subduction zone and underthrust Tibet. Owing to its low density, continental crust cannot be subducted deeply, so subduction and accompanying volcanism soon ended. However, India has continued moving northward relative to Asia. This continued movement is facilitated in part by southward thrusting of the Himalayas. Major uplift of the Himalayas was underway by 35 million years ago (fig. 11–13C), at which time the Main Central Thrust was active. This thrust transported rocks of the Tibetan accretionary-wedge and Indian continental shelf as well as Indian basement rocks.

Sometime between 20 and 10 million years ago (fig. 11–13D), this thrust stopped moving and a new, deeper fault, the Main Boundary Thrust, became active. This fault movement uplifted rocks of the first thrust sheet still more and also overthrust mountain-derived sediment on the Ganges Plain. Indian lithosphere was depressed by the added load of these thrust sheets, and both thrust sheets suffered considerable erosion. As a result of thrusting, many Himalayan peaks are capped with sedimentary rock that originated in the Tethys Sea. This indicates minimum vertical uplift, not counting eroded material, of at least 8 km.

Himalayan underthrusting was produced by India's northward movement against the Tibet Plateau, or conversely the advancing load of thick Tibetan crust caused overthrusting of the Himalayas. The latter point of view shows the similarity to ice-pushing (compare fig. 11–13 with previous diagrams of ice thrusting). Tibet fills the role of an ice sheet, whose forward movement imposed an increasing load on the substratum. The accretionary-wedge complex is analogous to ice-contact and proglacial sediment deposited during glacier advance. India represents pre-existing strata that were thrust in front of the advancing ice, and lithosphere depression occurred in response to both tectonic and glacial loading.

India's northward movement is also accommodated in another, quite different way by displacement along major transform faults to the east, north, and west (Tapponnier et al. 1982). North-south transform faults bound the eastern (Burma-Bay of Bengal) and western (Pakistan-Arabian Sea) sides of the Indian plate. These boundary faults are analogous to the lateral margins of certain ice-pushed hills, where tear faults, low ridges or elongated drumlins are developed parallel to ice movement and transverse to composite-ridges (Chapter 2). However, the mosaic of faults and rift zones that disrupts a vast region extending north of India, across China, as far as Lake Baikal in the Soviet Union does not have any glaciotectonic comparison.

The southern Appalachian and Ouachita Mountains of eastern North America were built during the Paleozoic by continental collisions involving ancestral North America and Gondwanaland. During the collisions, thick thrust sheets were pushed 100s of km onto the American continent. The thrust sheets include igneous and metamorphic rocks along with a large volume of sedimentary rock that originated on the former continental margin. Prior to thrusting, the sedimentary strata were porous, full of fluid, and contained abundant hydrated minerals.

Oliver (1986) speculated on the fate of fluids involved in Appalachian and Ouachita thrusting. Some of the fluid migrated up into the metamorphic interior of the mountains; some escaped to hot springs along the mountain front; some remained in decollements to facilitate thrusting; some may have descended into the basement; and some fluid was expelled into permeable sedimentary strata of the foreland beyond the mountains. These latter fluids carried heat, dissolved minerals and hydrocarbons into the mid-continent region with widespread geologic consequences. The zonation of coal metamorphism, hydrocarbons, and secondary mineralization shows definite relationship to the mountain belts (fig. 11–14).

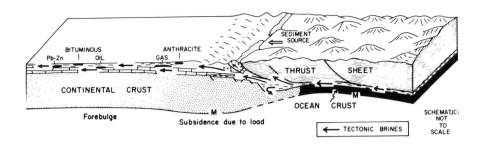

Fig. 11-14. Block diagram showing simplified thrusting of southern Appalachian or Ouachita Mountains. Note zonation of features associated with migration of tectonic brines into the mid-continent. Taken from Oliver (1986, fig. 2).

Oliver (1986) compared the action of the thrust sheet to a giant squeegee or roller that drives fluids ahead as it advances. This situation is analogous to the outer zone of glaciation, where ice sheets thrust poorly consolidated and saturated sedimentary strata. Ground water was forced to flow toward the ice margin. This fluid was driven out of compacted sediments as well as derived from subglacial melting and infiltration from the ice surface. Where trapped in confined aquifers or beneath permafrost or stagnant ice, ground water migrated into the proglacial region before escaping at springs. Being relatively cold and dilute, ground water emerging from beneath ice sheets had little chemical or metamorphic effects. However, its erosional and depositional effects could be striking (Gustavson and Boothroyd 1987).

Thrusting of ice-shoved hills is highly dependent on such high-pressure fluid within decollements, just as thrusting of mountains requires fluid support. From a mechanical point of view, then, creation of major thrust mountains along convergent plate boundaries is really no different from thrusting of ice-shoved hills.

BIBLIOGRAPHY

Aarseth, I. and Mangerud, J. 1974. Younger Dryas end moraines between Hardangerfjorden and Sognefjorden, western Norway. *Boreas* 3:3–22.
Aber, J.S. 1979. Kineto-stratigraphy at Hvideklint, Møn, Denmark and its regional significance. *Geological Society Denmark, Bulletin* 28:81–93.
Aber, J.S. 1982. Model for glaciotectonism. *Geological Society Denmark, Bulletin* 30:79–90.
Aber, J.S. 1985a. The character of glaciotectonism. *Geologie en Mijnbouw* 64:389–395.
Aber, J.S. 1985b. Definition and model for Kansan glaciation. *Ter-Qua Symposium Series*, vol. 1:53–60.
Aber, J.S. 1988a. *Structural geology exercises with glaciotectonic examples*. Hunter Textbooks, Winston-Salem, North Carolina, 140 p.
Aber, J.S. 1988b. Spectrum of constructional glaciotectonic landforms. *In* Goldthwait, R.P. and Matsch, C.L. (eds.), *Genetic classification of glacigenic deposits*, p. 281–292. A.A. Balkema, Rotterdam.
Aber, J.S. 1988c. Geomorphic and structural genesis of the Dirt Hills and Cactus Hills, Saskatchewan. *In* Fenton, M.M. (ed.), in press. A.A. Balkema, Rotterdam.
Aber, J.S. 1988d. West Atchison drift section. *Geological Society America, Centennial Field Guide, South-central section*, p. 5–10.
Aber, J.S. 1988e. Bibliography of glaciotectonic references. *In* Croot, D.G. (ed.), *Glaciotectonics forms and processes*, p. 195–212. A.A. Balkema, Rotterdam.
Aber, J.S. 1988f. Ice-shoved hills of Saskatchewan compared with Mississippi delta mudlumps--Implications for glaciotectonic models. *In* Croot, D.G. (ed.), *Glaciotectonics forms and processes*, p. 1–9. A.A. Balkema, Rotterdam.
Aber, J.S. 1988g. The glaciation of Kansas. *Heritage of the Great Plains* vol. 21, no. 4.
Aber, J.S. and Aarseth, I. 1988. Glaciotectonic structure and genesis of the Herdla Moraines, western Norway. *Norsk Geologisk Tidsskrift* 68:99–106.
Aber, J.S., Abdelsaheb, I., Nutter, B., Denne, J.E. and MacDonald, W.D. 1988. Composition, paleomagnetism, and age of the Kansas Drift. *Kansas Academy Science, Abstracts* 7:1.
Aber, J.S. and Lundqvist, J. 1988. Glaciotectonic structures in central Sweden and their significance for glacial theory. *Géographique Physique et Quaternaire* 42/3:315–323.
Adrielsson, L. 1984. Weichselian lithostratigraphy and glacial environments in the Ven-Gumslöv area, southern Sweden. *Lundqua Thesis* 14, Lund University, Sweden, 120 p.
Agricola, G. 1546. *De Natura Fossilium*. Basle.
Andersen, H.V. 1961. Genesis and paleontology of the Mississippi River mudlumps, Part II Foraminifera of the mudlumps, lower Mississippi River delta. *Louisiana Geological Survey, Geological Bulletin* 35, part II, 208 p.
Andersen, S.A. 1966. En fremstilling af Danmarks Kvartærgeologi. Anmeldelse og vurdering af S. Hansen: The Quaternary of Denmark. *In* Rankama, K. (ed.), *The Quaternary*, vol. 1, Interscience Publ. (1965). *Meddelelser Dansk Geologisk Forening* 16:214–233.
Andriashek, L.D. and Fenton, M.M. 1988. Quaternary stratigraphy and surficial geology, Sand River Map Sheet 73L, Alberta. *Terrain Sciences Department, Alberta Research Council, Bulletin*, in press.
Andriashek, L.D., Kathol, C.P., Fenton, M.M. and Root, J.D. 1979. Surficial geology Wabamun Lake NTS 83G. *Alberta Research Council Map*.
Andrews, D.E. 1980. Glacially thrust bed rock--an indication of late Wisconsin climate in western New York State. *Geology* 8:97–101.
Andrews, J.T. 1970. A geomorphological study of post-glacial uplift with particular reference to Arctic Canada. *Institute British Geographers, Special Publ.* No. 2, London, 156 p.
Anundsen, K. 1985. Changes in shore-level and ice-front positions in late Weichsel and Holocene, southern Norway. *Norsk Geografisk Tidsskrift* 39:205–225.

Babcock, E.A., Fenton, M.M. and Andriashek, L.D. 1978. Shear phenomena in ice-thrust gravels, central Alberta. *Canadian Journal Earth Sciences* 15:277–283.
Bally et al. 1966. Structure, seismic data, and orogenic solution of southern Canadian Rocky Mountains. *Bulletin Canadian Petroleum Geologists* 14:337–381
Banham, P.H. 1975. Glacitectonic structures: A general discussion with particular reference to the contorted drift of Norfolk. *In* Wright, A.E. and Moseley, F. (eds.), Ice ages: ancient and modern. *Geological Journal, Special Issue* 6:69–94.
Banham, P.H. 1977. Glacitectonics in till stratigraphy. *Boreas* 6:101–105.
Banham, P.H. 1988a. Thin-skinned glaciotectonic structures. *In* Croot, D.G. (ed.), *Glaciotectonics forms and processes.* p. 21–25. A.A. Balkema, Rotterdam.
Banham, P.H. 1988b. Polyphase glaciotectonic deformation in the Contorted Drift of Norfolk. *In* Croot, D.G. (ed.), *Glaciotectonics forms and processes,* p. 27–32. A.A. Balkema, Rotterdam.
Barbour, E.H. 1913. A minor phenomenon of the glacial drift in Nebraska. *Nebraska Geological Survey,* vol. 4, part 9:161–164.
Bell, R. 1874. Report on the country between Red River and the South Saskatchewan. *Geological Survey Canada, Report of Progress,* 1873–74, p. 66–93.
Ber, A. 1987. Glaciotectonic deformation of glacial landforms and deposits in the Suwalki Lakeland (NE Poland). *In* Meer, J.J.M. van der (ed.), *Tills and glaciotectonics,* p.135–143. A.A. Balkema, Rotterdam.
Berg, M.W. van den and Beets, D.J. 1987. Saalian glacial deposits and morphology in the Netherlands. *In* Meer, J.J.M. van der (ed.), *Tills and glaciotectonics,* p. 235–251. A.A. Balkema, Rotterdam.
Berthelsen, A. 1973. Weichselian ice advances and drift successions in Denmark. *Bulletin Geol. Instit. Univ. Uppsala, n.s.* 5:21–29.
Berthelsen, A. 1974. Nogle forekomster af intrusivt moræneler i NØ-Sjælland. *Dansk Geologisk Forening, Årsskrift for* 1973:118–131.
Berthelsen, A. 1975. Geologi på Røsnæs. *Varv, Ekskursionsfører* 3, 78 p.
Berthelsen, A. 1978. The methodology of kineto-stratigraphy as applied to glacial geology. *Geological Society Denmark, Bulletin* 27, Special Issue, p. 25–38.
Berthelsen, A. 1979. Recumbent folds and boudinage structures formed by sub-glacial shear: An example of gravity tectonics. *In* Linden, W.J.M. van der (ed.), Van Bemmelen and his search for harmony. *Geologie en Mijnbouw* 58:253–260.
Berthelsen, A., Konradi, P. and Petersen, K.S. 1977. Kvartære lagfølger og strukturer i Vestmøns klinter. *Dansk Geologisk Forening, Årsskrift for* 1976:93–99.
Billings, M.P. 1972. *Structural Geology,* 3rd edition. Prentice-Hall, New Jersey, 606 p.
Birkenmajer, K. 1987. Pleistocene-Holocene boundary events in South Spitsbergen. *Norsk Polarinstitutt, late Cenozoic paleoenvironments and geology of the Arctic, Program.* April 27–29.
Bluemle, J.P. 1970. Anomalous hills and associated depressions in central North Dakota. *Geological Society America, Abstracts with Prog.* 2:325–326.
Bluemle, J.P. 1981. Geology of Sheridan County, North Dakota. *North Dakota Geological Survey, Bulletin* 75, part 1, 59 p.
Bluemle, J.P. and Clayton, L. 1984. Large-scale glacial thrusting and related processes in North Dakota. *Boreas* 13:279–299.
Bouchard, M. 1974. *Geologie de depots de L'Ile Herschel, Territorie du Yukon.* Unpubl. Master's thesis, University Montreal, Montreal.
Boyer and Elliott 1982. Thrust systems. *American Assoc. Petroleum Geologists, Bulletin* 41:2603–2676.
Brodzikowski, K. and Loon, A.J. van 1985. Inventory of deformational structures as a tool for unraveling the Quaternary geology of glaciated areas. *Boreas* 14:175–188.
Byers, A.R. 1959. Deformation of the Whitemud and Eastend Formations near Claybank, Saskatchewan. *Transactions Royal Society Canada* 53, series 3, sect. 4:1–11.
Carlson, C.G. and Freers, T.F. 1975. Geology of Benson and Pierce Counties, North Dakota. *North Dakota Geological Survey, Bulletin* 59, part 1.

Carrigy, M.A. 1970. Proposed revision of the boundaries of the Paskapoo Formation in the Alberta Plains. *Bulletin Canadian Petroleum Geology* 18:156–165.
Chamberlin, T.C. 1886. The rock-scorings of the great ice invasions. *United States Geological Survey, 7th Annual Report* III:147–248.
Chapple, W.M. 1978. Mechanics of thin-skinned fold-and-thrust belts. *Geological Society America, Bulletin* 89:1189–1198.
Christiansen, E.A. 1956. Glacial geology of the Moose Mountain area Saskatchewan. *Saskatchewan Department Mineral Resources, Report* 21.
Christiansen, E.A. 1961. Geology and ground-water resources of the Regina Area Saskatchewan. *Saskatchewan Research Council, Geology Division, Report* 2.
Christiansen, E.A. 1971a. Tills in southern Saskatchewan, Canada. *In* Goldthwait, R.P. (ed.), *Till/a symposium*, p. 167–183. Ohio State University Press, Columbus.
Christiansen, E.A. 1971b. Geology and groundwater resources of the Melville Area (62K, L) Saskatchewan. *Saskatchewan Research Council, Geology Division, Map* No. 12.
Christiansen, E.A. 1979. The Wisconsinan deglaciation of southern Saskatchewan and adjacent areas. *Canadian Journal Earth Sciences* 16:913–938.
Christiansen, E.A. and Whitaker, S.H. 1976. Glacial thrusting of drift and bedrock. *In* Legget, R.F. (ed.), Glacial till. *Royal Society Canada, Special Publication* 12:121–130.
Clayton, L. and Moran, S.R. 1974. A glacial process-form model. *In* Coates, D.R. (ed.), *Glacial geomorphology*, p. 89–119. SUNY-Binghamton Publications in Geomorphology, Binghamton, New York.
Clayton, L., Moran, S.R. and Bluemle, J.P. 1980. Explanatory text to accompany the Geologic Map of North Dakota. *North Dakota Geological Survey, Report of Investigation* No. 69.
Coates, D.R. 1974. Reappraisal of the glaciated Appalachian Plateau. *In* Coates, D.R. (ed.), *Glacial geomorphology*, p. 203–243. SUNY-Binghamton Publications in Geomorphology, Binghamton, New York.
Coleman, J.M. 1988. Dynamic changes and processes in the Mississippi River delta. *Geological Society America, Bulletin* 100:999–1015.
Croot, D.G. 1978. *The depositional landforms and sediments produced by two surging glaciers.* Unpubl. Ph.D. thesis, University of Aberdeen, U.K.
Croot, D.G. 1987. Glacio-tectonic structures: A mesoscale model of thin-skinned thrust sheets? *Journal Structural Geology* 9:797–808.
Croot, D.G. (ed.) 1988a. *Glaciotectonics forms and processes*, A.A. Balkema, Rotterdam, 212 p.
Croot, D.G. 1988b. Morphological, structural and mechanical analysis of neoglacial ice-pushed ridges in Iceland. *In* Croot, D. G. (ed.), *Glaciotectonics forms and processes.* p. 33–47. A. A. Balkema, Rotterdam.
Croot, D.G. 1988c. Glaciotectonics and surging glaciers: A correlation based on Vestspitsbergen, Svalbard, Norway. *In* Croot, D.G. (ed.), *Glaciotectonics forms and processes.* p. 49–61. A. A. Balkema, Rotterdam.
Dahlstrom, C.D.A. 1970. Structural geology in the eastern margin of the Canadian Rocky Mountains. *Bulletin Canadian Petroleum Geology* 18/3.
Dahlstrom, C.D.A. 1977. Structural geology in the eastern margin of the Canadian Rocky Mountains. *29th Annual Field Conference Wyoming Geol. Assoc.* p. 407–439.
Dellwig, L.F. and Baldwin, A.D. 1965. Ice-push deformation in northeastern Kansas. *Kansas Geological Survey, Bulletin* 175, part 2, 16 p.
Dort, W. Jr. 1966. Nebraskan and Kansan stades: Complexity and importance. *Science* 154:771–772.
Dort, W. Jr. 1985. Field evidence for more than two early Pleistocene glaciations in the central plains. *Ter-Qua Symposium Series* 1:41–51.
Dredge, L.A. and Grant, D.R. 1987. Glacial deformation of bedrock and sediment, Magdalen Islands and Nova Scotia, Canada: Evidence for a regional grounded ice sheet. *In* Meer, J.J.M. van der (ed.), *Tills and glaciotectonics*, p. 183–195. A.A. Balkema, Rotterdam.
Dreeszen, V.H. and Burchett, R.R. 1971. Buried valleys in the lower part of the Missouri River basin. *Kansas Geological Survey, Special Distribution Publication* 53:21–25.

Drewry, D. 1986. *Glacial geologic processes*. Edward Arnold, London, 276 p.
Drozdowski, E. 1981. Pre-Eemian push-moraines in the lower Vistula region, northern Poland. *In* Ehlers, J. and Zandstra, J.G. (eds.), Glacigenic deposits in the southwest parts of the Scandinavian Icesheet. *Mededelingen Rijks Geologische Dienst* 34–1/11:57–61.
Dyke, A.S. and Prest, V.K. 1987a. Late Wisconsinan and Holocene history of the Laurentide Ice Sheet. *Géographie physique et Quaternaire* XLI:237–263.
Dyke, A.S. and Prest, V.K. 1987b. Wisconsinan and Holocene retreat of the Laurentide Ice Sheet. *Geological Survey Canada, Map* 1702A, scale = 1:5,000,000.
Dylik, J. 1961. The Łódź region. *VI INQUA Congress, Guidebook of excursion,* Łódź.
Ehlers, J. 1978. Fine gravel analysis after the Dutch method tested out on Ristinge Klint, Denmark. *Geological Society Denmark, Bulletin* 27:157–165.
Elliott, D. 1976. The energy balance and deformation mechanisms of thrust sheets. *Philosophical Transactions Royal Society London* 283:289–312.
Elverhøi, A., Lønne, Ø. and Seland, R. 1983. Glaciomarine sedimentation in a modern fjord environment, Spitsbergen. *Polar Research, n.s.* 1:127–149.
Erdmann, E. 1873. Iaktagelser öfver moränbildningar och deraf betäckta skiktade jordarter i Skåne. *Geologiska Föreningens i Stockholm, Förhandlingar* 2:13–24.
Eybergen, F.A. 1987. Glacier snout dynamics and contemporary push moraine formation at the Turtmannglacier, Wallis, Switzerland. *In* Meer, J.J.M van der (ed.), *Tills and glaciotectonics*, p. 217–231. A.A. Balkema, Rotterdam.
Fenton, M.M. 1984. Quaternary stratigraphy of the Canadian Prairies. *In* Fulton, R.J. (ed.), Quaternary of Canada--A Canadian contribution to IGCP Project 24. *Geological Survey Canada, Paper* 84–10:57–68.
Fenton, M.M. and Andriashek, L.D. 1983. Surficial geology Sand River area, Alberta (NTS 73L). *Alberta Geological Survey, Map*, scale = 1:250,000.
Fenton, M.M., Langenberg, C.W., Jones, C.E., Trudell, M.R., Pawlowicz, J.G., Tapics, J.A. and Nikols, D.J. 1985. Tour of the Highvale open pit coal mine. Petroleum Society of CIM and Canadian Society Petroleum Geologists, Guidebook, *Alberta Research Council, Open File Report* 1985-7, 55 p.
Fenton, M.M., Moell, C.E., Pawlowicz, J.G., Sterenberg, G., Trudell, M.R. and Moran, S.R. 1983. *Highwall stability project, Highvale Mine study report, December 1983*. Unpublished report for TransAlta Utilities by Alberta Geological Survey, Alberta Research Council, 70 p.
Fenton, M.M., Moran, S.R., Teller, J.T. and Clayton, L. 1983. Quaternary stratigraphy and history in the southern part of the Lake Agassiz basin. *In* Teller, J.T. and Clayton, L. (eds.), Glacial Lake Agassiz. *Geological Association Canada, Special Paper* 26:49–74.
Fenton, M.M. and Pawlowicz, J.G. in prep. Importance of glaciotectonism to coal exploration and development. *Alberta Research Council, Terrain Science Dept., Open File Report*.
Fenton, M.M., Trudell, M.R., Pawlowicz, J.G., Jones, C.E., Moran, S.R. and Nikols, D.J. 1986. Glaciotectonic deformation and geotechnical stability in open pit coal mining. *In* Singhal, R.K. (ed.), *Geotechnical stability in surface mining*, p. 225–234. A.A. Balkema, Rotterdam.
Fernlund, J.M.R. 1988. The Halland coastal moraines: Are they end moraines or glaciotectonic ridges? *In* Croot, D.G. (ed.), *Glaciotectonics forms and processes*, p. 77–90. A.A. Balkema, Rotterdam.
Flint, R.F. 1971. *Glacial and Quaternary geology*. J. Wiley and Sons, New York, 892 p.
Fraser, F.J., McLearn, F.H., Russell, L.S., Warren, P.S. and Wickenden, R.T.D. 1935. Geology of Saskatchewan. *Geological Survey Canada, Memoir* 176, 137 p.
Galon, R. 1961. North Poland, area of the last glaciation. *VI INQUA Congress, Guidebook of excursion*. Łódź.
Gans, W. de, Groot, T. de and Zwaan, H. 1987. The Amsterdam basin, a case study of a glacial basin in the Netherlands. *In* Meer, J.J.M. van der (ed.), *Tills and glaciotectonics*, p. 205–216. A.A. Balkema, Rotterdam.
Garwood, E.J. and Gregory, J.W. 1898. Contributions to the glacial geology of Spitsbergen. *Quarterly Journal Geological Soc. London* 5A:197–227.

Gijssel, K. van 1987. A lithostratigraphic and glaciotectonic reconstruction of the Lamstedt Moraine, Lower Saxony (FRG). In Meer, J.J.M. van der (ed.), Tills and glaciotectonics, p. 145–155. A.A. Balkema, Rotterdam.
Goldthwait, R.P. 1971. Introduction to till, today. In Goldthwait, R.P. (ed.), Till/a symposium, p. 3–26. Ohio State University Press, Columbus.
Graham, R.H. 1978. Wrench faults, arcuate fold patterns and deformation in the southern French Alps. Proceedings Geologists' Association 89:125–142.
Gripp, K. 1929. Glaciologische und geologische Ergebnisse der Hamburgischen Spitzbergen-Expedition 1927. Abh. naturwiss. Ver. Hamburg XXII:147–249.
Gripp, K. and Todtman, E.M. 1925. Die Endmoräne des Green-Bay Gletschers auf Spitsbergen, eine studie zum Verständnis norddeutscher Diluvial-Gebilde. Mitt. geogr. Ges. Hamburg 37:45–75.
Gry, H. 1940. De istektoniske forhold i moleromraadet. Meddelelser Dansk Geologisk Forening 9:586–627.
Gry, H. 1965. Furs geologi. Fur Museum, Dansk Natur-Dansk Skole, Årsskrift for 1964, 55 p.
Gry, H. 1979. Beskrivelse til geologisk kort over Danmark, Kortbladet Løgstor, Kvartære aflejringer. Danmarks Geologiske Undersøgelse, I række 26.
Gustavson, T.C. and Boothroyd, J.C. 1987. A depositional model for outwash, sediment sources, and hydrologic characteristics, Malaspina Glacier, Alaska: A modern analog of the southeastern margin of the Laurentide Ice Sheet. Geological Society America, Bulletin 99:187–200.
Haarsted, V. 1956. De kvartærgeologiske og geomorfologiske forhold på Møn. Meddelelser Dansk Geologisk Forening 13:124–126.
Haldorsen, S. and Sørensen, R. 1987. Distribution of tills in southeastern Norway. In Meer, J.J.M. van der (ed.), Tills and glaciotectonics, p. 31–38. A.A. Balkema, Rotterdam.
Hall, R.E., Poppe, L.J. and Ferrebee, W.M. 1980. A stratigraphic test well, Martha's Vineyard, Massachusetts. United States Geological Survey, Bulletin 1488, 19 p.
Hansen, S. 1930. Om forekomsten af glacialflager af palæocæn mergel på Sjælland. Danmarks Geologiske Undersøgelse, IV række 2(7), 22 p.
Harding, T.P. and Lowell, J.D. 1979. Structural styles, their plate-tectonic habitats and hydrocarbon traps in petroleum provinces. American Assoc. Petroleum Geologists, Bulletin 63:1016–1058.
Harland, W.B., Herod, K.N. and Krinsley, D.H. 1966. The definition and identification of tills and tillites. Earth Science Review 2:225–256.
Harris, L.D. and Milici, R.C. 1977. Characteristics of thin-skinned style of deformation in the southern Appalachians and potential hydrocarbon traps. United States Geological Survey, Profession Paper 1018, 40 p.
Hart, J.K. 1987. The genesis of the North East Norfolk Drift. Unpubl. Ph.D. thesis, University of East Anglia, U.K.
Hillefors, Å. 1985. Deep-weathered rock in western Sweden. Fennia 163/2:293–301.
Hintze, V. 1937. Moens Klints geologi. C.A. Reitzels, Copenhagen, 410 p. (edited posthumously by E.L. Mertz and V. Nordmann).
Hirvas, H., Korpela, K. and Kujansuu, R. 1981. Weichselian in Finland before 15,000 BP. Boreas 10:423–431.
Hobbs, B.E., Means, W.D. and Williams, P.F. 1976. An outline of structural geology. J. Wiley and Sons, New York, 571 p.
Hollick, A. 1894. Dislocations in certain portions of the Atlantic coastal plain strata and their probable causes. Transactions New York Academy Sciences 14:8–20.
Holmström, L. 1904. Öfversikt af den glaciala afslipningen i Sydskandinavien. Geologiska Föreningens i Stockholm Förhandlingar 26:241–316, 365–432.
Holst, N.O. 1903. Om skriftkritan i tullstorpstrakten och de båda moräner, in hvilka den är inbäddad. Sveriges Geologiska Undersökning, C194.
Holst, N.O. 1911. Beskrifning till kartbladet Börringekloster. Sveriges Geologiska Undersökning, Aa138.

Holter, M.E., Yurko, J.R. and Chu, M. 1975. Geology and coal reserves of the Ardley Coal Zone of central Alberta. *Alberta Research Council, Report* 75-7, 41 p.

Hopkins, O.B. 1923. Some structural features of the plains area of Alberta caused by Pleistocene glaciation. *Geological Society America, Bulletin* 34:419–430.

Horberg, L. and Anderson, R.C. 1956. Bedrock topography and Pleistocene glacial lobes in central United States. *Journal Geology* 64:101–116.

Houmark-Nielsen, M. 1981. Glacialstratigrafi i Danmark øst for Hovedopholdslinien. *Dansk Geologisk Forening, Årsskrift for* 1980:61–76.

Houmark-Nielsen, M. 1987. Pleistocene stratigraphy and glacial history of the central part of Denmark. *Geological Society Denmark, Bulletin* 36:1–189.

Houmark-Nielsen, M. 1988. Glaciotectonic unconformities in Pleistocene stratigraphy as evidence for the behavior of former Scandinavian icesheets. *In* Croot, D.G. (ed.), *Glaciotectonics forms and processes*, p. 91–99. A.A. Balkema, Rotterdam.

Howe, W.B. 1968. Guidebook to Pleistocene and Pennsylvanian formations in the St. Joseph area, Missouri. *Missouri Geological Survey and Water Resources, Association Missouri Geologists, 15th annual field trip and meeting*, 45 p.

Hubbert, M.K. and Rubey, W.W. 1959. Role of fluid pressure in mechanics of overthrust faulting. *Geological Society America, Bulletin* 70:115–166.

Humlum, O. 1978. A large till wedge in Denmark: implications for the subglacial thermal regime. *Geological Society Denmark, Bulletin* 27:63–71.

Humlum, O. 1983. Dannelsen af en disloceret randmoræne ved en avancerende isrand, Höfdabrekkujökull, Island. *Dansk Geologisk Forening, Årsskrift for* 1982:11–26.

Irish, E.J.W. 1970. The Edmonton Group of south-central Alberta. *Bulletin Canadian Petroleum Geology* 18:125–155.

Jahn, A. 1950. Nowe dane o położeniu kry jurajskiej w Lukowie (New facts concerning the ice transported blocks of the Jurassic at Luków). *Annales Societatis Geologorum Poloniae* 19:372–385, Kraków.

Jahn, A. 1956. Wyzyna Lubelska, rzeźba i czwartorzed. (Geomorphology and Quaternary history of Lublin Plateau.) *Państw. Wyd. Nauk.*, Warszawa, 453 p.

Jelgersma, S. and Breeuwer, J.B. 1975. Toelichting bij de kaart glaciale verschijnselen gedurende het Saalian, 1:600,000. *In* Zagwijn, W.H. and Staalduinen, C.J. (eds.), *Toelichting bij geologische overzichtskaarten van Nederland*, p. 93–103. Rijks Geologische Dienst, Haarlem.

Jessen, A. 1931. Lønstrup Klint. *Danmarks Geologiske Undersøgelse, II række* 49, 142 p.

Johnstrup, F. 1874. Ueber die Lagerungsverhaltnisse und die Hebungs-phänomene in den Kreidefelsen auf Moen und Rügen. *Zeit. deutsch. geol. Ges.* 1874:533–585.

Jong, J.D. de 1952. On the structure of the pre-glacial Pleistocene of the Archemerberg (Prov. of Overijsel, Netherlands). *Geologie en Mijnbouw* 14:86.

Jong, J.D. de 1967. The Quaternary of the Netherlands. *In* Rankama, K. (ed.), *The Quaternary*, vol. 2:301–426. J. Wiley, New York.

Kalin, M. 1971. The active push moraine of the Thompson glacier. *Axel Heiberg Island Research Reports*, no. 4, McGill University, Montreal.

Kamb, B., Raymond, C.F., Harrison, W.D., Englehardt, H., Eschelmayer, K.A., Humphrey, N., Brugman, M.M. and Pfeffer, T. 1985. Glacier surge mechanism, 1981–1983 surge of Variegated Glacier, Alaska. *Science* 227:468–479.

Karczewski, A. 1984. Geomorphology of the Hornsund Fiord area Spitsbergen: Commentary to the map. Institute Geophysics, *Polish Academy Sciences*, Warszawa, 26 p.

Kaye, C.A. 1964a. Outline of Pleistocene geology of Martha's Vineyard, Massachusetts. *United States Geological Survey, Professional Paper* 501-C:134–139.

Kaye, C.A. 1964b. Illinoian and early Wisconsin moraines of Martha's Vineyard, Massachusetts. *United States Geological Survey, Professional Paper* 501-C:140–143.

Kaye, C.A. 1980. Geologic profile of Gay Head Cliff, Martha's Vineyard, Massachusetts. *United States Geological Survey, Open-file Report* 80–148.

Klassen, R.A. 1982. Glaciotectonic thrust plates, Bylot Island, District of Franklin. *Geological Survey Canada, Current Research, part A, Paper* 82–1A:369–373.

Klassen, R.W. 1975. Quaternary geology and geomorphology of Assiniboine and Qu'Appelle Valleys of Manitoba and Saskatchewan. *Geological Survey Canada, Bulletin* 228, 61 p.

Klassen, R.W. 1979. Pleistocene geology and geomorphology of the Riding Mountain and Duck Mountain areas, Manitoba-Saskatchewan. *Geological Survey Canada, Memoir* 396, 52 p.

Königsson, L.K. and Linde, L.A. 1977. Glaciotectonically disturbed sediments at Rönnerrum on the island of Öland. *Geologiska Föreningens i Stockholm, Förhandlingar* 99:68–72.

Konradi, P.B. 1973. Foraminiferas in some Danish glacial deposits. *Bulletin Geol. Instit. Univ. Uppsala, n.s.* 5:173–175.

Krahn, J., Johnson, R.F., Fredlund, D.G. and Clifton, A.W. 1979. A highway failure in Cretaceous sediments at Maymount, Saskatchewan. *Canadian Geotechnical Journal* 16:703–715.

Kulhawy, F.H. 1975. Stress deformation properties of rock and rock discontinuities. *Engineering Geology* 9:327–350.

Kupsch, W.O. 1955. Drumlins with jointed boulders near Dollard, Saskatchewan. *Geological Society America, Bulletin* 66:327–338.

Kupsch, W.O. 1962. Ice-thrust ridges in western Canada. *Journal Geology* 70:582–594.

Lagerlund, E. 1987. An alternative Weichselian glaciation model, with special reference to the glacial history of Skåne, South Sweden. *Boreas* 16:433–459.

Lagerlund, E., Knutsson, G., Åmark, M., Hebrand, M., Jönsson, L.-O., Karlgren, B., Kristiansson, J., Möller, P., Robison, J.M., Sandgren, P., Terne, T. and Waldemarsson, D. 1983. The deglaciation pattern and dynamics in South Sweden, A preliminary report. *Lundqua Report*, vol. 24, Lund, Sweden, 8 p.

Lammerson, P.R. and Dellwig, L.F. 1957. Deformation by ice push of lithified sediments in south-central Iowa. *Journal Geology* 65:546–550.

Lamplugh, G.W. 1911. On the shelly moraine of the Sefstromglacier and other Spitsbergen phenomena illustrative of the British glacial conditions. *Proceedings Yorkshire Geological Society* 17:216–241.

Lavrushin, Y.A. 1971. Dynamische Fazies und Subfazies der Grundmoräne. *Zeit. Angew. Geol.* 17:337–343.

Lemoine, M. 1973. About gravity gliding in the western Alps. *In* Jong, K.A. de and Scholten, R. (eds.), *Gravity tectonics*, p. 201–216. J. Wiley and Sons, New York.

Lewiński, J. and Różycki, S.Z. 1929. Dwa profile geologiczne przez Warszawe. *Sprawozd. Tow. Nauk. Warsz.* 22:30–50.

Lowell, J.D. 1985. *Structural styles in petroleum exploration*. Oil and Gas Consultants International Publication, Tulsa, 460 p.

Lundqvist, J. 1967. Submoräna sediment i Jämtlands Län. *Sveriges Geologiska Undersökning*, C618, 267 p.

Lundqvist, J. 1985. Glaciations and till or tillite genesis: Examples from Pleistocene glacial drift in central Sweden. *Palaeogeography, Palaeoclimatology, Palaeoecology* 51:389–395.

Lyell, C. 1863. *The geological evidences of the antiquity of man*, 3rd ed. John Murray, London.

Maarleveld, G.C. 1953. Standen van het landijs in Nederland. *Boor en Spade* 4:95–105.

Mackay, J.R. 1959. Glacier ice-thrust features of the Yukon coast. *Geographical Bulletin* 13:5–21.

Mackay, J.R. and Mathews, W.H. 1964. The role of permafrost in ice-thrusting. *Journal Geology* 72:378–380.

Mackay, J.R., Rampton, V.N. and Fyles, J.G. 1972. Relic Pleistocene permafrost, western Arctic, Canada. *Science* 176:1321–1323.

Mackay, J.R. and Stager, J.K. 1966. Thick tilted beds of segregated ice, Mackenzie delta area, N.W.T. *Biuletyn Peryglacjalny* 15:39–43.

Madsen, V. 1916. Ristinge Klint. *Danmarks Geologiske Undersøgelse, IV række* 1(2), 32 p.

Madsen, V., Nordmann, V. and Hartz, N. 1908. Eem-zonerne. Studier over Cyprinaleret og andre Eem-aflejringer i Danmark, Nord-Tyskland og Holland. *Danmarks Geologisk Undersøgelse, II række* 17, 302 p.

Mangerud, J. and Skreden, S.A. 1972. Fossil ice wedges and ground wedges in sediments below till at Voss, western Norway. *Norsk Geologisk Tidsskrift* 52:73–96.
Mangerud, J., Sønstegaard, E., Sejrup, H.-P. and Haldorsen, S. 1981. A continuous Eemian-early Weichselian sequence containing pollen and marine fossils at Fjøsanger, western Norway. *Boreas* 10:137–208.
Mathews, W.H. and MacKay, J.R. 1960. Deformation of soils by glacier ice and the influence of pore pressures and permafrost. *Transactions Royal Society Canada* 54, series 3, section 4:27–36.
McGinn, R.A. and Giles, T.R. 1987. The McFadden Valley-Polonia Trench spillway system, Riding Mountain, Manitoba. *Internat. Union Quaternary Research, XII Internat. Congress, Prog. with Abstracts*, p. 223.
Meer, J.J.M. van der (editor) 1987. *Tills and glaciotectonics*. A.A. Balkema, Rotterdam, 270 p.
Meier, M.F. and Post, A.S. 1969. What are glacier surges. *Canadian Journal Earth Sciences* 6(4) II, 807–818.
Merrill, F.J.H. 1886a. On the geology of Long Island. *Annals New York Academy Sciences* 3:341–364.
Merrill, F.J.H. 1886b. On some dynamic effects of the ice-sheet. *Proceedings American Assoc. Advancement Science* 35:228–229.
Meyer, K.-D. 1987. Ground and end moraines in Lower Saxony. *In* Meer, J.J.M. van der (ed.), *Tills and glaciotectonics*, p. 197–204. A.A. Balkema, Rotterdam.
Moell, C.E., Pawlowicz, J.G., Fenton, M.M., Trudell, M.R., Jones, C.E. and Sterenberg, G. 1985. *Highwall stability project, Highvale Mine study report for 1984*. Unpublished report for TransAlta Utilities by Terrain Sciences Dept., Alberta Research Council, 140 p.
Molnar, P. 1986. The structure of mountain ranges. *Scientific American* 255/1:70–79.
Moran, S.R. 1971. Glaciotectonic structures in drift. *In* Goldthwait, R.P. (ed.), *Till/a symposium*, p. 127–148. Ohio State University Press, Columbus.
Moran, S.R., Clayton, L., Hooke, R.LeB., Fenton, M.M. and Andriashek, L.D. 1980. Glacier-bed landforms of the Prairie region of North America. *Journal Glaciology* 25:457–476.
Morgan, J.P. 1961. Genesis and paleontology of the Mississippi River mudlumps, Part I Mudlumps at the mouths of the Mississippi River. *Louisiana Geological Survey, Geological Bulletin* 35, part I, 116 p.
Morgan, J.P., Coleman, J.M. and Gagliano, S.M. 1968. Mudlumps: diapiric structures in Mississippi Delta sediments. *In* Braunstein, J. and O'Brien, G.D. (eds.), Diapirism and diapirs. *American Assoc. Petroleum Geologists, Memoir* 8:145–161.
Mulugeta, G. and Koyi, H. 1987. Three-dimensional geometry and kinematics of experimental piggyback thrusting. *Geology* 15: 1052–1056.
Occhietti, S. 1973. Les structures et déformations engendrées par les glaciers--Essai de mise au point. *Revue Géographique de Montréal* 27:365–380.
Oldale, R.N. 1980. Pleistocene stratigraphy of Nantucket, Martha's Vineyard, the Elizabeth Islands, and Cape Cod, Massachusetts. *In* Larson, G.J. and Stone, B.D. (eds.), *Late Wisconsin glaciation of New England*, p. 1–34. Kendall/Hunt, Dubuque, Iowa.
Oldale, R.N. and O'Hara, C.J. 1984. Glaciotectonic origin of the Massachusetts coastal end moraines and a fluctuating late Wisconsinan ice margin. *Geological Society America, Bulletin* 95:61–74.
Oliver, J. 1986. Fluids expelled tectonically from orogenic belts: Their role in hydrocarbon migration and other geologic phenomena. *Geology* 14:99–102.
Parizek, R.P. 1964. Geology of the Willow Bunch Lake Area (72-H) Saskatchewan. *Saskatchewan Research Council, Geology Division, Report* 4, 46 p.
Paterson, W.S.B. 1981. *The physics of glaciers*. Pergamon, Oxford, U.K. 380 p.
Peake, N.B. and Hancock, J.M. 1961. The upper Cretaceous of Norfolk. *Trans. Norfolk Norwich Natural Society* 19:293–339.
Pedersen, G.K. and Surlyk, F. 1983. The Fur Formation, a late Paleocene ash-bearing diatomite from northern Denmark. *Geological Society Denmark, Bulletin* 32:43–65.

Pedersen, S.A.S. 1986. Videregående undersøgelser af sandkiler på Fur. *Danmarks Geologiske Undersøgelse, Intern rapport* nr. 32 (1986).
Pedersen, S.A.S. 1987. Comparative studies of gravity in tectonic Quaternary sediments and sedimentary rocks related to fold belts. *In* Jones, M.E. and Preston, R.M.F. (eds.), Deformation of sediments and sedimentary rocks, *Geological Society, Special Publication* 29:165–180.
Pedersen, S.S. and Petersen, K.S. 1985. Sandkiler i moler på Fur. *Danmarks Geologiske Undersøgelse, Intern rapport* nr. 32 (1985).
Pedersen, S.A.S. and Petersen, K.S. 1988. Sand filled frost wedges in glaciotectonically deformed mo-clay on the island of Fur, Denmark. *In* Croot, D.G. (ed.), *Glaciotectonics forms and processes*. p. 185–190. A.A. Balkema, Rotterdam.
Pedersen, S.A.S., Petersen, K.S. and Rasmussen, L.A. 1988. Observations on glaciodynamic structures at the Main Stationary Line in western Jutland, Denmark. *In* Croot, D.G. (ed.), *Glaciotectonics forms and processes*, p. 177–183. A.A. Balkema, Rotterdam.
Perry, W.J., Roeder, D.H. and Lageson, D.R. 1984. North American thrust faulted terranes. *American Association Petroleum Geologists, Reprint* 27.
Petersen, K.S. 1978. Applications of glaciotectonic analysis in the geological mapping of Denmark. *Danmarks Geologiske Undersøgelse, Årbog* 1977:53–61.
Prest, V.K. 1976. Quaternary geology. *In* Douglas, R.J.W. (ed.), Geology and economic minerals of Canada. *Geological Survey Canada, Economic Geology Report* No. 1, part B:675–764.
Prest, V.K. 1983. Canada's heritage of glacial features. *Geological Survey Canada, Miscellaneous Report* 28, 119 p.
Prest, V.K. 1984. Late Wisconsinan glacial complex. *Geological Survey Canada, Map* 1584A, *in Paper* 84–10.
Prest, V.K., Grant, D.R. and Rampton, V.N. 1967. Glacial Map of Canada. *Geological Survey Canada, Map* 1253A, scale = 1:5,000,000.
Puggaard, C. 1851. Møns Klint section. Reproduced in *International Geological Congress XXI, Session Norden* (1960), *Guidebook* I.
Rampton, V.N. 1982. Quaternary geology of the Yukon Coastal Plain. *Geological Survey Canada, Bulletin* 317, 49 p.
Rich, J.L. 1934. Mechanics of low-angle overthrust faulting as illustrated by Cumberland thrust block, Virginia, Kentucky and Tennessee. *American Assoc. Petroleum Geologists, Bulletin* 18/12:1584–1596.
Ringberg, B. 1980. Beskrivning till Jordartskartan Malmö SO (Description to the Quaternary map Malmö SO). *Sveriges Geologiska Undersökning*, Ae38.
Ringberg, B. 1983. Till stratigraphy and glacial rafts of chalk at Kvarnby, southern Sweden. *In* Ehlers, J. (ed.), *Glacial deposits in North-west Europe*, p. 151–154. A.A. Balkema, Rotterdam.
Ringberg, B. 1988. Late Weichselian geology of southernmost Sweden. *Boreas* 17:243–263.
Ringberg, B., Holland, B. and Miller, U. 1984. Till stratigraphy and provenance of the glacial chalk rafts at Kvarnby and Ängdala, southern Sweden. *Striae* 20:79–90.
Roeder, D.L. 1983. Hydrocarbons and geodynamics of fold-thrust belts. *Rocky Mountain Assoc. Geologists, Continuing Education short course notes*, 216 p.
Rotnicki, K. 1976. The theoretical basis for and a model of the origin of glaciotectonic deformations. *Quaestiones Geographicae* 3:103–139.
Ruegg, G.H.J. 1981. Ice pushed lower and middle Pleistocene deposits near Rhenen (Kwintelooijen): sedimentary-structural and lithological/granulometrical investigations. *In* Ruegg, G.H.J. and Zandstra, J.G. (eds.), Geology and archaeology of Pleistocene deposits in the ice-pushed ridge near Rhenen and Veenendaal. *Mededelingen Rijks Geologische Dienst* 35–2/7:165–177.
Ruegg, G.H.J. and Zandstra, J.G. (eds.) 1981. Geology and archaeology of Pleistocene deposits in the ice-pushed ridge near Rhenen and Veenendaal. *Mededelingen Rijks Geologische Dienst* 35–2/7:163–268.
Ruszczyńska-Szenajch, H. 1976. Glacitektoniczne depresje i kry lodowcowe na tle budowy geologicznej południowo-wschodniego Mazowsza i południowego Podlasia.

(Glacitectonic depressions and glacial rafts in mid-eastern Poland). *Studia Geologica Polonica* 50:1–106. Warszawa.

Ruszczyńska-Szenajch, H. 1978. Glacitectonic origin of some lake-basins in areas of Pleistocene glaciations. *Polskie Archiwum Hydrobiologii* 25:373–381.

Ruszczyńska-Szenajch, H. 1985. Origin and age of the large-scale glaciotectonic structures in central and eastern Poland. *Annales Societatis Geologorum Poloniae* 55:307–332, Kraków.

Ruszczyńska-Szenajch, H. 1986. The origin of glacial rafts: Detachment, transport, deposition. *INQUA's Working Group on Glacial Tectonics, Field Meeting, Abstracts*, Oct. 3–5, 1986.

Ruszczyńska-Szenajch, H. 1987. The origin of glacial rafts: Detachment, transport, deposition. *Boreas* 16:101–112.

Ruszczyńska-Szenajch, H. 1988. Glaciotectonics and its relationship to other glaciogenic processes. In Croot, D.G. (ed.), *Glaciotectonics forms and processes*, p. 191–193. A.A. Balkema, Rotterdam.

Sardeson, F.W. 1905. A particular case of glacial erosion. *Journal Geology* 13:351–357.

Sardeson, F.W. 1906. The folding of subjacent strata by glacial action. *Journal Geology* 14:226–232.

Sauer, E.K. 1978. The engineering significance of glacier ice-thrusting. *Canadian Geotechnical Journal* 15:457–472.

Schafer, J.P. and Hartshorn, J.H. 1965. The Quaternary of New England. In Wright, H.E. Jr. and Frey, D.G. (eds.), *The Quaternary of the United States*, p. 113–127. Princeton Univ. Press, Princeton, New Jersey.

Shaler, N.S. 1888. Geology of Martha's Vineyard. *United States Geological Survey, Report for* 1886, vol. 3:297–363.

Shaler, N.S. 1898. Geology of the Cape Cod district. *United States Geological Survey, Report for* 1896–97, part 2:497–593.

Sharp, M. 1982. *A comparison of the landforms and sedimentary sequences produced by surging and non-surging glaciers in Iceland*. Unpubl. Ph.D. thesis, University of Aberdeen, U.K.

Siddans, A.W.B. 1979. Arcuate fold and thrust patterns in the Subalpine chains of southeast France. *Journal Structural Geology* 1:117–126.

Siddans, A.W.B. 1984. Thrust tectonics--A mechanistic view from West and Central Alps. *Tectonophysics* 104:257–281.

Sirkin, L. 1976. Block Island, Rhode Island: Evidence of fluctuation of the late Pleistocene ice margin. *Geological Society America, Bulletin* 87:574–580.

Sirkin, L. 1980. Wisconsinan glaciation of Long Island, New York to Block Island, Rhode Island. In Larson, G.J. and Stone, B.D. (eds.), *Late Wisconsin glaciation of New England*, p. 35–59. Kendall/Hunt, Dubuque, Iowa.

Sjørring, S. 1981. The Weichselian till stratigraphy in the southern part of Denmark. *Quaternary Studies in Poland* 3:103–109.

Sjørring, S. 1983. Ristinge Klint. In Ehlers, J. (ed.), *Glacial deposits in North-west Europe*, p. 219–226. A.A. Balkema, Rotterdam.

Sjørring, S. (editor) 1985. INQUA and IGCP field meeting in Denmark 1981. *Geological Society Denmark, Bulletin* 34:1.

Sjørring, S., Nielsen, P.E., Frederiksen, J.K., Hegner, J., Hyde, G., Jensen, J.B., Morgensen, A. and Vortisch, W. 1982. Observationer fra Ristinge Klint, felt- og laboratorieundersøgelser. *Dansk Geologisk Forening, Årsskrift for* 1981:135–149.

Slater, G. 1926. Glacial tectonics as reflected in disturbed drift deposits. *Geologists' Association Proceedings* 37:392–400.

Slater, G. 1927a. The structure of the disturbed deposits in the lower part of the Gipping Valley near Ipswich. *Geologists' Association Proceedings* 38:157–182.

Slater, G. 1927b. The structure of the disturbed deposits of the Hadleigh Road area, Ipswich. *Geologists' Association Proceedings* 38:183–261.

Slater, G. 1927c. The structure of the disturbed deposits of Møens Klint, Denmark. *Transactions Royal Society Edinburgh* 55, part 2:289–302.

Slater, G. 1927d. The disturbed glacial deposits in the neighborhood of Lönstrup, near Hjörring, north Denmark. *Transactions Royal Society Edinburgh* 55, part 2:303–315.
Slater, G. 1927e. Structure of the Mud Buttes and Tit Hills in Alberta. *Geological Society America, Bulletin* 38:721–730.
Slater, G. 1929. The structure of the drumlins exposed on the south shore of Lake Ontario. *New York State Museum, Bulletin* 281:3–19.
Slater, G. 1931. The structure of the Bride Moraine, Isle of Man. *Proceedings Liverpool Geological Society* 14:184–196.
Smed, P. 1962. Studier over den fynske øgruppes glaciale landskabsformer. *Meddelelser Dansk Geologisk Forening* 15:1–74.
Sollid, J.-L. and Reite, A. 1983. The last glaciation and deglaciation of central Norway. In Ehlers, J. (ed.), *Glacial deposits in North-west Europe*, p. 41–59. A.A. Balkema, Rotterdam.
Stalker, A.MacS. 1973a. Surficial geology of the Drumheller Area, Alberta. *Geological Survey Canada, Memoir* 370.
Stalker, A.MacS. 1973b. The large interdrift bedrock blocks of the Canadian Prairies. *Geological Survey Canada, Paper* 75–1A:421–422.
Stalker, A.MacS. 1976. Megablocks, or the enormous erratics of the Albertan Prairies. *Geological Survey Canada, Paper* 76–1C:185–188.
Stephan, H.-J. 1985. Deformations striking parallel to glacier movement as a problem in reconstructing its direction. *Geological Society Denmark, Bulletin* 34:47–53.
Stone, B.D. and Koteff, C. 1979. A late Wisconsinan ice readvance near Manchester, New Hampshire. *American Journal Science* 279:590–601.
Surlyk, F. 1971. Skrivekridtklinterne på Møn. In Hansen, M. and Poulsen, V. (eds.), Geologi på øerne. *Varv, Ekskursionsfører* 2:5–23.
Sønstegaard, E. 1979. Glaciotectonic deformation structures in unconsolidated sediments at Os, south of Bergen. *Norsk Geologisk Tidsskrift* 59:223–228.
Tapics, J.A. 1984. The Highvale Mine--Developments in the next ten years. *86th Annual Meeting of CIM–1984, Paper No.* 78, 17 p.
Tapponnier, P., Peltzer, G., Le Dain, A.Y., Armijo, R. and Cobbold, P. 1982. Propagating extrusion tectonics in Asia: New insights from simple experiments with plasticine. *Geology* 10:611–616.
Teller, J.T. 1983. Jean de Charpentier 1786–1855. *Biobibliographical Studies* 7:17–22, Mansell Publ. Co., London.
Teller, J.T., Moran, S.R. and Clayton, L. 1980. The Wisconsinan deglaciation of southern Saskatchewan and adjacent areas: Discussion. *Canadian Journal Earth Sciences* 17:539–541.
Ter-Borch, N. and Tychsen, J. 1987. Kalkoverfladens struktur (Structural map of top chalk group). *Skov- og Naturstyrelsen, Havbundsundersøgelsen og Dansk Olie- og Gasproduktion A/S*, scale = 1:500,000.
Torell, O. 1872. Undersökningar öfver istiden del I. Aftryck ur *Öfversigt af Kungliga Vetenskapsakademiens Förhandlingar* 1872, P.A. Nordstedt och Söner, Stockholm, 44 p.
Torell, O. 1873. Undersökningar öfver istiden del II. Skandinaviska landisens utsräckning under isperioden. *Öfversigt af Kungliga Vetenskapsakademiens Förhandlingar* 1873, no. 1:47–64.
Upham, W. 1899. Glacial history of the New England islands, Cape Cod, and Long Island. *American Geologist* 24:79–92.
Vorren, T.O. 1979. Weichselian ice movements, sediments and stratigraphy on Hardangervidda, South Norway. *Norges Geologiske Undersøkelse* 350 (Bull. 50), 117 p.
Washburn, A.L. 1980. *Geocryology--A survey of periglacial processes and environments*. J. Wiley and Sons, New York, 406 p.
Wateren, D.F.M. van der 1981. Glacial tectonics at the Kwintelooijen sandpit, Rhenen, the Netherlands. In Ruegg, G.H.J. and Zandstra, J.G. (eds.), Geology and archaeology of Pleistocene deposits in the ice-pushed ridge near Rhenen and Veenendaal. *Mededelingen Rijks Geologische Dienst* 35–2/7:252–268.

Wateren, D.F.M. van der 1985. A model of glacial tectonics, applied to the ice-pushed ridges in the central Netherlands. *Geological Society Denmark, Bulletin* 34:55–74.

Wateren, D. van der 1987. Structural geology and sedimentology of the Dammer Berge push moraine, FRG. *In* Meer, J.J.M. van der (ed.), *Tills and glaciotectonics*, p. 157–182. A.A. Balkema, Rotterdam.

Wee, M.W. ter 1962. The Saalian glaciation in the Netherlands. *Meded. Geol. Stichting NS* 15:57–77.

Wee, M.W. ter 1983. The Saalian glaciation in the northern Netherlands. *In* Ehlers, J. (ed.), *Glacial deposits in North-west Europe*, p. 405–412. A.A. Balkema, Rotterdam.

Weertman, J. 1961. Mechanism for the formation of inner moraines found near the edge of cold ice caps and ice sheets. *Journal Glaciology* 3:965–978.

Welsted, J. and Young, H.R. 1980. Geology and origin of the Brandon Hills, southwest Manitoba. *Canadian Journal Earth Sciences* 17:942–951.

White, W.A. 1972. Deep erosion by continental ice sheets. *Geological Society America, Bulletin* 83:1037–1056.

Woodworth, J.B. 1897. Unconformities of Martha's Vineyard and of Block Island. *Geological Society America, Bulletin* 8:197–212.

Woodworth, J.B. and Wigglesworth, E. 1934. Geography and geology of the region including Cape Cod, the Elizabeth Islands, Nantucket, Martha's Vineyard, No Mans Land, and Block Island. *Memoirs Museum Comparative Zoology, Harvard College*, 322 p.

Zandstra, J.G. 1981. Petrology and lithostratigraphy of ice-pushed lower and middle Pleistocene deposits at Rhenen (Kwintelooijen). *In* Ruegg, G.H.J. and Zandstra, J.G. (eds.), Geology and archaeology of Pleistocene deposits in the ice-pushed ridge near Rhenen and Veenendaal. *Mededelingen Rijks Geologische Dienst* 35–2/7:178–191.

INDEX

Note: This index includes text material only for Chapters 1–11; figures may be located in relation to geography or subject.

acoustic velocity, sediment 113
Agassiz, Louis 1
Asia 180
   Himalaya Mountains 180, 181
   Tethys Sea 180, 181
   Tibet Plateau 180, 181

Banham, Peter H. 4
Berthelsen, Asger 4, 6
Bluemle, John L. 5
Byers, A.R. 4

Canada, Alberta 1, 4, 5, 16–20, 37, 121, 136
   Cooking Lake megablock 91
   Driftwood Bend megablock 97
   Highvale Coal Mine 121–127
   Oldman River 96–98
   southern Alberta megablocks 94–98
   Sundance Power Plant 121
   Wolf Lake 16–20
Canada, Bylot Island 47–49
Canada, general 1, 4–5, 37, 47, 119, 139, 143, 162
   Alberta Plain 38, 98
   Manitoba Plain 49
   Saskatchewan Plain 38, 49, 94
Canada, Magdalen Islands 139
Canada, Manitoba 49, 54, 136
   Brandon Hills 49–54
Canada, Nova Scotia 139
Canada, Saskatchewan 4–5, 34, 43, 121, 136, 143
   Ardill end moraine 41–42
   Dirt Hills/Cactus Hills 4, 33, 37–42, 146
   Esterhazy megablock 94–95
   Maymount 127–129
   Qu'Appelle Valley 94
Canada, Yukon 136
   Coastal Plain 5, 21
   Herschel Basin 21
   Herschel Island 16, 21–24
Cenozoic 91, 136, 180
Charpentier, Jean de 1
China 181
Christiansen, Earl 5

Clayton, Lee 5
Cretaceous 13, 34–35, 39, 46, 73, 78–83, 94–95, 97–99, 122, 136, 146
   Ardley Coal Zone 136
   Bearpaw Formation 39, 96
   Campanian 85
   Eastend Formation 39, 42
   Grand Rapids Formation 91
   Hell Creek Formation 43, 45
   Judith River Formation 128
   Maastrichtian 35, 85, 99
   Odanah Member 94–95
   Paskapoo Formation 136
   Ravenscrag Formation 39, 42
   Riding Mountain Formation 49, 94–95
   Scollard Member 136
   Whitemud Formation 39, 42
crust/lithosphere, depression 7, 24, 116, 141, 181

Denmark 3–4, 73, 77, 85, 117–118, 137, 151–152
   Bornholm 140
   Flade Klit, Mors 30–31
   Fur 121, 129–134
   Fyn 74
   Hanklit, Mors 31
   Hundborg, Jylland 13
   Hvideklint, Møn 73, 83–88
   Høje Møn 35–37
   Limfjord 3, 31, 130, 134
   Lønstrup Klint, Jylland 2
   Manhøj quarry, Fur 131
   Møn 88, 99, 104
   Møns Klint 1, 31–37, 78, 85
   Ristinge Klint, Langeland 73–77
   Systofte, Falster 106, 116–118
   Ærø 73–76

Eemian 35, 74, 76, 86, 101
Europe 3, 135, 137–138
   Alps 1, 173, 180
   Baltic Sea 99, 102, 137, 152
   Caledonian Mountains 138
   Central European Plain 137
   North Sea 137

Norwegian Sea 137
Fennoscandian Shield 138
Fennoscandian Border Zone 99, 140
Jura Mountains 31
Scandinavia 140

Fennoscandian Ice Sheet 4, 135, 140
Finland 138, 140
fjord
   Spitsbergen 49, 67
   western Norway 112, 116
foraminifera 35, 86, 172
France 174

Germany 4, 19, 138
   Rügen 1, 99, 138
glacial tectonics, first use 2
glacial theory 1, 6
glaciation
   drumlin 16, 19, 71, 73–74, 136, 139, 148, 168, 181
   end moraine 29, 49, 77, 83, 136–138, 149
   erratic 1, 22, 91, 135, 150
   esker 45, 50, 53, 138, 144, 148
   forebulge 175
   kame 13, 22, 117
   moraine 1, 7, 10, 22, 42, 50, 67, 99, 112, 128
   nunatak 42, 82
   Rogen moraine 140
   sandur 26, 47, 49, 55, 59, 61, 65
   spillway 41–42, 56–57, 148
   striation/groove/abrasion 1, 91, 101, 107, 113, 116, 135, 150
   till fabric 7, 101, 107, 109, 117, 150
   tunnel valley 57, 148
glacier ice 6, 7, 104
   block movement 19
   calving 116
   lobe geometry 143, 150
   surge/surging 25, 49, 58, 63, 66, 69, 109, 116
glaciodynamic 7
glaciofluvial 123, 132–134
glaciokarstic 7
glaciolacustrine 109, 123
glaciomarine 67, 112
glaciostatic 7
glaciotectonic 7
glaciotectonic analogs 169
   comparison to mountains 10, 31, 169
   convergent plate boundary 169, 179–182
   delta mudlump 169–173
   fluid migration 181–182
   gravity sliding/gliding 175
   gravity spreading 175–178
   pushing from the rear 177–178
   thin-skinned thrusting 169, 173–179
glaciotectonic deformation 119, 155–169, 178
   brittle 9, 104
   cohesion/cohesive strength 158, 160–162
   competence 41–42, 63
   Coulomb principle 161
   domainal deformation 149–150
   ductile 9, 65, 104
   extra-domainal deformation 149–150
   fundamental cause 155
   internal friction 158, 160
   penetrative 148, 150
   piggyback thrusting 164, 168, 172
   scale model 164–168
glaciotectonic landforms 9–12
   composite-ridges, definition 29
   composite-ridges, general 29, 50, 58, 65, 71, 82, 142, 146, 149, 163, 165, 173, 179–181
   composite-ridges, large 10, 29–47, 49, 82, 129, 169, 173
   composite-ridges, small 10, 24, 26, 47–69, 121, 129, 131, 167, 169
   cupola-hill 10, 24, 28, 71–88, 117, 129, 142
   cupola-hill morphology 71, 148
   depression/basin 10, 13, 16–19, 26–28, 45–47, 54, 56–58, 63, 91, 119, 144–146, 155
   glaciotectonic landscape 148–149
   hill-hole pair 10, 13–28, 71, 129, 142, 144
   hill-hole pair, definition 13
   hill-hole pair, morphology 13–16
   ice-shoved hill 4, 10, 16, 20, 29, 39, 54, 57, 71, 140, 143–146, 159, 163–165, 168, 182
   ice-shoved hill, definition 10
   irregular hills 71
   kuppelbakke 71
   materials 11
   megablock 10, 91–102, 142, 159, 162–164, 168
   megablock, definition 10
   megablock plain 96
   morphostructural region 35
   murdlin 16
   pseudo-moraine 10
   push-moraine 3, 47–49, 61, 67, 69, 166, 168

INDEX

push-moraine, definition 10
Stauchrücken 10
Stauchendmoränen 10
Stauchmoränen 10
stuuwmorenen 10
stuuwwallen 10
trellis drainage 23
glaciotectonic structures 8–9
  allochthonous 8, 30
  anticline/syncline 23, 35, 61, 72, 87, 108, 114–115, 132–134
  apophysis 51, 106
  autochthonous 8
  breccia/brecciation 8–9, 56, 85, 91, 95, 101, 128, 132
  decollement 28, 56, 58, 61, 63, 77, 95, 133, 148, 159, 169, 178, 182
  diapir 9, 103–116, 142
  drag fold 76, 115, 117–118, 165
  endiamict 7
  exodiamict 7
  fault/faulting 8–9, 50–51, 85, 91, 108, 123–124, 126, 128–131
  fissure 9, 105, 133–134, 164
  floe 28, 35, 71, 77–78, 83, 85–88, 166–167
  floe, definition 10
  fold/folding 8–9, 23, 29, 38–39, 45, 56, 63, 72, 79, 82, 87, 91, 98, 108, 114–115, 119, 123–124, 126, 128–134
  foliation 51, 76, 106, 114
  fracture/fracturing 9, 91, 123, 125–126, 134, 165
  groove 106
  identification of 8
  intrusion 9, 103–104, 108–109, 114–116
  kink/kinking 51, 165
  melange 61, 86, 96, 98
  mylonite 95
  normal fault 51, 56, 61, 63, 87, 134, 160, 164
  overthrust fault 82
  plug 104, 109
  raft 86, 91, 94, 96, 98–102, 142
  raft, definition 10
  scale 31, 35, 39, 47, 49, 56, 74, 76, 78, 82
  scale, definition 10
  shear plane/zone/band 24, 56, 60, 87, 91, 98, 123–127
  sill 104, 109
  slickensides 8, 24, 56, 95, 98, 106, 123, 128–129
  slide/slump 125–128, 164–165
  tear fault 17, 19, 181
  thrust block/sheet 17, 29, 47, 56, 61–63, 128, 144, 163–165
  thrust fault 23, 39, 47, 51, 61–65, 72, 74, 76, 79, 86–87, 91, 106, 108, 114–115, 119, 121–124, 134, 159–164
  underthrust 165, 175
  vein 105–106
  wedge 103, 105–106, 116–118, 133–134
  xenolith 106
glaciotectonics 2, 7
glaciotectonics, applied 119–134
  coal mine/mining 119–120
  drift prospecting 119
  highwall failure 121–127
  highway construction 121, 127–129
  mine planning/operation 119–122, 127, 131, 134
  mineral exploration 119
  soils mapping 119
  soil salinity 119
glaciotectonics, distribution 135–153
  continent-scale 135–142
  inner zone 135, 140–141
  intermediate zone 135, 138–140
  lobate model 143, 145–149
  outer zone 135–138, 142
  regional 142–147
  transition belt 139–140
  zonal model 135–142
glaciotectonism, definition 6–7
glacitectonics 2
Gondwanaland 181
ground water 46, 123–129, 142, 144, 161, 182
  pressure 63, 73, 83, 104, 126, 148, 162, 168
  aquifer 46, 73, 83, 143–145, 148, 162, 182
  conductivity 63, 127
  spring/fountain 63, 83, 162, 182
Gry, Helge 3

Holocene
  glaciation 34, 49, 59, 68
  Little Ice Ages 47, 67

iceberg drifting 1, 7
Iceland 16, 49, 166
  Eyjabakkajökull 24–28, 59–65
  Höfdabrekkujökull 59, 166–168
  Kviarjökull 59
  Mýrdalsjökull 166

Skeidararjökull 59
Vatnajökull 25, 59
Illinoian glaciation 80
India 180–181
　Main Boundary Thrust 181
　Main Central Thrust 180
INQUA 6
Ireland 137

Jessen, Axel 2
Jurassic 91, 101, 140

Kansan glaciation 91, 106–110
　Atchison Formation 107–109
　Dakota lobe 110
　Kansas Drift 106–107, 110
　Lower Kansas Till 107–109
　Minnesota lobe 110
　Upper Kansas Till 107–109
kineto-stratigraphy 4, 149–153
　drift unit, definition 149
　principle 150
Kupsch, Walter 4
landslide, general 1, 79, 127, 169

Laurentide Ice Sheet 4, 53, 83, 135, 140, 141
lithosphere, definition 180
Lyell, Charles 1

Mackay, J.R. 5
Mathews, W.H. 5
Mesozoic 91, 136
Miocene 80
Moran, Steven R. 5

Netherlands 3, 54, 56–57, 137, 146
　Kwintelooijen 54–55
　Utrecht Ridge 54–58, 146
North America 4, 135–136, 139–141, 181
　Appalachian Mountains 139, 173, 181
　Atlantic Coastal Plain 78, 83, 136
　Canadian Shield 139
　Rocky Mountains 1, 31, 136, 173
　Central Lowland 139
　Great Lakes 139
　Great Plains 5, 20, 107, 136, 143, 146
　Missouri Coteau 38, 40, 43, 143
　Ouachita Mountains 181
Norway 110–112, 116, 138, 140
　Herdla Moraines 106, 110–116
　Voss 102
Norway, Svalbard 65
　Chomjakovbreen 67
　Hornbreen 67

Mendelejevbreen 67
Spitsbergen 3, 49, 65–69, 116
Storbreen 67
Svalisbreen 67
Treskelen 67–69

Paleocene 13, 30, 122, 130
　Danian 99
　Fur Formation 130–131
Paleolithic artifacts 54
Paleozoic 78, 86, 91, 101, 107, 139–140, 181
Pelukian transgression 22
Pennsylvanian 91
　Elmont Limestone Member 91
　Tarkio Limestone Member 91
　Willard Shale 91
　Zeandale Formation 91
permafrost 5, 9, 11, 21, 24, 47–48, 58, 77, 95, 98, 106, 109, 133–134, 142, 144, 148, 162, 182
　ice wedge 106, 133
　patterned ground 49
Permian 139
piezometer 123, 128
Pleistocene 16, 21, 25, 34, 49, 80, 82, 107, 135, 143, 147, 164
　Aquinnah conglomerate 80
　Bow Valley gravel 98
　Coleharbor Group 43
　Cromer Tills 104–105
　Kedichem Formation 55–56, 58
　Urk Formation 55, 57
Poland 1, 4, 13, 138, 146
　Lukow rafts 91
　Suwalki Lakeland 149
pre-Quaternary strata 8, 11, 21, 29, 47
Precambrian 78, 140, 149
　Sioux Quartzite 110
pressure/stress 155–164, 177
　angle of friction 160, 166–167
　differential loading 56, 58, 172–173
　glaciodynamic pressure 155, 158–159
　glaciostatic pressure 155–156, 159, 164
　hydrostatic 24, 77, 161–162, 168–170
　intergranular 161
　lateral 56, 58, 156–159, 162, 172–173
　lithostatic 161–162
　normal stress 159–160, 177
　Poisson's ratio 156
　shear stress/strength 56, 127–128, 155, 159–163, 169

Quaternary strata/geology 4, 11, 37, 47, 49, 54, 73–74, 83, 91, 119, 138–139, 149, 155

radiocarbon (C-14) dates 47, 53, 68, 82
rock/sediment types
  ash 31, 129–130
  bedding plane 8, 47, 161–162
  bentonite 39, 97–98, 123–124
  chalk 9, 31, 34–35, 73, 83–88, 98–102, 104–105, 138
  chert/flint 35, 39, 101, 108
  clay/silt 9, 21, 24, 26, 35, 55, 61–65, 67, 76–77, 83, 96, 103–106, 109, 113–114, 117, 119, 123, 169–170, 172
  claystone 41, 94–95, 130–131, 133, 161–162
  coal 96–98, 119–124, 181
  diatomite 30, 129–130
  dolostone 39
  glauconite/greensand 80
  hydrocarbon 181
  kaolinite 39, 80
  lignite 38–39, 41, 43, 82, 161–162
  limestone 9, 86, 99, 101, 107, 140
  moler 129–134
  montmorillonite 130
  mudstone 38–39, 42–43, 123–124, 126
  peat 21, 55, 61
  quartzite 39
  quick-clay 103
  sand/gravel 9, 21, 26, 35, 45, 47, 50, 55, 61, 63–65, 67 80, 95, 98, 105, 107–109, 113, 117, 131–134, 144, 159, 161, 164, 166–167, 169, 171–173
  sandstone 9, 39, 41–43, 46, 96–98, 119, 123, 125–126, 128, 139–140
  shale 9, 38–39, 43, 49, 94–98, 107, 119, 123, 128, 140, 161
  siltstone 43, 94, 128
  slate 139
  soil/turf/loam 55, 61–65
  stratified drift 49–51, 53, 75, 83, 85, 87, 107, 117–118, 140, 149
  tephra 26, 61–63, 130
  till 7, 22, 35, 43, 46–47, 50–53, 55, 57, 61, 71–72, 75–76, 79, 85–88, 92, 95–102, 105–109, 113, 116–119, 123, 128, 134, 136, 138, 140, 148–149, 168

Saalian glaciation 54–57, 74–75, 88, 146, 148
  Drente Formation 55
  Gelderse Vallei lobe 54–56
Sangamon 22, 80
Slater, George 1

soft-sediment deformation 104, 109, 114
South America
  Andes Mountains 180
Soviet Union 137–138, 140
  Lake Baikal 181
squeeze box 164
Sweden 98, 102, 137–138, 140–141
  Alnarp Graben 99
  Kvarnby 94, 98–102
  Skåne 1, 98, 102
  Tornquist Line 140
  Ven 72
  Öland 140
  Öresund 72, 102
Switzerland 1, 31
  Turtmannglacier 168

Tertiary 73, 78–82, 101, 136, 146–147
  Cannonball Formation 43, 45

United Kingdom
  England 1, 137
  Isle of Man 2
  Norfolk 1, 4, 104
United States, Alaska 22
  Malaspina Glacier 83
United States, general 1, 5, 41, 51, 143
United States, Iowa 110, 136, 140
United States, Kansas 106–107, 110, 140
  Topeka megablock 91
United States, Massachusetts 82, 136
  Gay Head, Martha's Vineyard 73, 77–82, 146
  Vineyard moraine 78–80
United States, Minnesota 1
  Coteau des Prairies 110
United States, Mississippi
  Mississippi delta 169–173
  South Pass distributary 169, 172
United States, Missouri 110, 140
United States, Nebraska 110, 140
United States, New England 1, 77–78, 83
United States, New Hampshire 139
United States, New York 82, 140
  Valley Heads Moraine 140
United States, North Dakota 5, 10, 110, 136, 143–144
  Anamoose 144–145
  Antelope Hills 13
  Denhoff Hills 46
  Prophets Mountains 33, 42–46
United States, Rhode Island 82
United States, Wisconsin 5, 140

Weichselian glaciation 10, 30, 35–37, 72, 135, 151

Asnæs interstade 77
Bælthav advance 77, 151
East-Jylland phase 77
ice-dome model 152–153
ice-lobe model 152–153
Low Baltic advance 102
Main Weichselian advances 36, 76–77, 88, 102, 151
Norwegian advance 151
Old Baltic advance 36, 76, 171
Røsnæs interstade 77
Storebælt readvance 37, 88
Young Baltic advances 37, 76–77, 88, 118, 152
Young Baltic lobe 76, 151
Younger Dryas glaciation 112–113, 115–116
Weichselian stratigraphy 4, 35, 74–76, 86, 88, 102, 140
   Bælthav Till 37, 76
   East Jylland Till 37, 76
   Kvarnby Till 99–101
   Mid Danish Till 37, 76–77, 88
   North Sjælland Till 37, 88
   Ristinge Klint Till 36, 76, 88
   S. Sallerup Till 101
   Sunnanå Till 101
WGGT 6
Wisconsin glaciation 10, 22, 42, 46, 78, 80, 95, 135, 172
   Assiniboine sublobe 51
   Avonlea tongue 41
   Buckland glaciation 22, 24
   Cape Cod Bay lobe 78
   Cold Lake glaciation 20
   Galilee tongue 41
   Lostwood glaciation 41, 51
   Manhasset Formation 80
   Marchand phase 51
   Menemsha tongue 82
   Montauk Till 80, 82
   Narragansett-Buzzards Bay lobe 78, 82
   Primrose lobe 20
   Spring Valley tongue 41
   Weyburn lobe 41

GLACIOLOGY AND QUATERNARY GEOLOGY

1. V. V. Bogorodsky, C. R. Bentley and P. E. Gudmandsen, Radioglaciology. 1985. ISBN 90-277-1893-8.
2. I. A. Zotikov, The Thermophysics of Glaciers. 1986. ISBN 90-277-2163-7.
3. V. V. Bogorodsky, V. P. Gavrilo and O. A. Nedoshivin, Ice Destruction. 1987. ISBN 90-277-2229-3.
4. C. J. van der Veen and J. Oerlemans (eds.), Dynamics of the West Antarctic Ice Sheet. 1987. ISBN 90-277-2370-2.
5. J. S. Aber, D. G. Croot and M. M. Fenton, Glaciotectonic Landforms and Structures. 1989. ISBN 0-7923-0100-5.
6. J. Oerlemans (ed.), Glacier Fluctuations and Climatic Change. 1989. ISBN 0-7923-0110-2.